Measurement of Construction Work

Measurement of Construction Work

Volume 2
Second Edition

Chris Wilcox FRICS, FIQS, AMBIM

and

John A. Snape ARICS, AMBIM

George Godwin Limited
The book publishing subsidiary
of the Builder Group

© Chris Wilcox and John A. Snape 1980

First published in Great Britain in 1972 by George Godwin Limited under the title *Worked Examples in Measurement of Construction Work* Volume 2

Second edition 1980

George Godwin Limited
The book publishing subsidiary
of the Builder Group
1-3 Pemberton Row, Fleet Street
London EC4

CIP Data

Wilcox, Chris
    Measurement of construction work.
    Vol.2.—2nd ed.
    1. Building—Estimates
    I. Title   II. Snape, John A   III. Worked
    examples in measurement of construction work
    624'.1   TH435

ISBN 0-7114-5511-2

Printed and bound by
Tonbridge Printers, Tonbridge, Kent

Contents

	List of plates	vi
	Preface	vii
I	Finishings	1
II	Windows	29
III	Doors	72
IV	Woodwork-Staircases and Fittings	121
	Appendix A Abbreviations	160
	Appendix B Joinery Labours	162
	Appendix C Schedules BS 1192	167

Volume 1 of this series covers Introduction, Principles of Measurement and Description, Simple Foundations, Structural Walls, Roofs, Floors, Partitions.

List of Plates

Plate No.		Page
1	Internal Finishings 'A'	9
2	Internal Finishings 'B'	21
3	External Finishings	27
4	Standard Steel Window	37
5	Stock Pattern Casement Window	45
6	Casement Window	55
7	Pivoted Casement	63
8	Sash Window	71
9	Framed Ledged and Braced Door	81
10	Panelled Door	91
11	Flush Door	99
12	Glazed Door and Sidelight	109
13	Flush Doors 'A' and Glazed Doors 'B'	120
14	Straight Flight Staircase	127
15	Dog Leg Staircase	133
16	Open Newel Staircase	141
17	Stock Pattern Units	145
18	Simple Counter	149
19	Locker Unit	153
20	Unframed Second Fixings	159

Preface

This book is complementary to Volume I of the series.

The principal aim of both books is to demonstrate a logical and systematic approach to the measurement of construction work by the use of a wide range of worked examples.

For many years examiners have complained that students do not adopt a logical sequence of measurement and cannot write adequate descriptions of the work involved. The author's own experience fully supports this contention and they have therefore placed considerable emphasis on these aspects of the work.

The text of this second edition has been revised to comply with the requirements of the 6th edition of the Standard Method of Measurement of Building Works. This edition of the Standard Method is accompanied by a Practice Manual and the student is recommended to make reference to this as an aid to the correct application of the rules of measurement.

The text otherwise follows closely the format of the first edition. This was intended primarily for use by quantity surveying students and by students on other advanced course (Institute of Building, City and Guilds and others). Technician Education Council (TEC) courses, now introduced, require a knowledge of measurement at a very early stage. The authors consider the presentation of the wide range of examples, with fully detailed drawings, to be of considerable value to students at this level.

The first edition was published under the title *Worked Examples in Measurement of Construction Work.* However, both authors and publishers felt that this did not adequately reflect the full scope of the work and the title has therefore been revised.

As with the previous edition the group system of measurement is recommended and has been used in all examples. One complete chapter has been devoted to each group and each chapter opens with a description of a logical and systematic approach to the measurement of work in that group. The importance of the approach has been further emphasised by including a fully detailed approach at the commencement of each of the worked examples.

The instruction in the important technique of writing descriptions, as provided in Volume I demonstrated that the technique may be quickly acquired by the application of four simple rules. These rules have been applied to all descriptions in the worked examples and SMM phraseology has been used throughout.

In order to maintain a realistic presentation appropriate to current practice, this book has been printed in A4 size and the papers ruled in accordance with BS 3327:1970 'Specification for Stationery for Quantity Surveying'. The various schedules and drawings comply with BS 1192:1969 'Building Drawing Practice'. The drawings have been produced to recommended scales. The drawings include the constructional details so necessary to students who must now study quantities and building construction simultaneously. The dimensions have been handwritten since this is the only realistic form of presentation of this work.

The commentary which accompanies each example includes references to the appropriate SMM Rules, which have not been reproduced since it is essential that the student makes full and frequent reference to the SMM in order to obtain a thorough understanding of the Rules and their application.

The authors consider that the demonstration and discussion which takes place in the lecture room are of prime importance: but their experience proves that the learning process is considerably aided by a comprehensive text, used in conjunction with the lectures, and available to the student for detailed study and permanent reference.

Both the publishers and the authors would be interested to hear from lecturers and students who have any comments or suggestions which might contribute to further editions of this book.

CHRIS WILCOX
JOHN SNAPE

Chapter I

Finishings

Approach

(1)   Divide large buildings into sections of manageable size. Where a large project includes several buildings measure each as a separate unit.

(2)   Compare the finishings schedule with the drawings to ensure that all rooms have been included. When a schedule is not provided the taker off is generally advised to prepare his own. The preferred form of schedule recommended by BS 1192 'Building Drawing Practice' is illustrated on Plate No. 2.

(3)   Measure direct from the schedule using the drawings only to provide room dimensions. Rooms with identical finishings should be grouped together.

(4)   A standard sequence of measurement should be adopted:-
  e.g. Ceiling finishings
     Floor finishings
     Wall finishings

(5)   Measure finishings over all openings, recesses, projections and similar features which can be more easily dealt with by separate adjustment.

Measurement of Finishings

(1)   Finishings shall be classified as Internal work or External work and given under an appropriate heading (SMM T3).

(2)   Ceilings and floors often have the same area and may be measured together.

(3)   When beams project below ceilings the finishings to the sides and soffits take their height classification from the height of the adjoining ceiling (SMM T3). Work to sides and soffits of attached beams shall be regarded as work to the abutting ceiling. Work to sides, soffits or tops of isolated beams (grouped together) shall be measured separately. Beams where the finishings are different from the abutting ceiling shall be deemed to be isolated (SMM T5). Each face must be considered separately, since work not exceeding 300 mm wide shall be so described (SMM T4).

(4)   The total floor finish usually comprises a screeded, floated or trowelled bed and the finishing material. The thickness of individual finishings varies considerably, so that on a level base the thickness of the bed must be adjusted in order to maintain a level surface. For example, with a total floor finish of 50 mm, a 25 mm thick wood block requires a 25 mm thick floated bed and a 3 mm thick PVC tile a 47 mm trowelled bed.

(5)   Plaster behind wood skirtings should be dealt with as work to walls disregarding any grounds (SMM T3). Where the plaster does not extend behind the skirting it should not be measured.

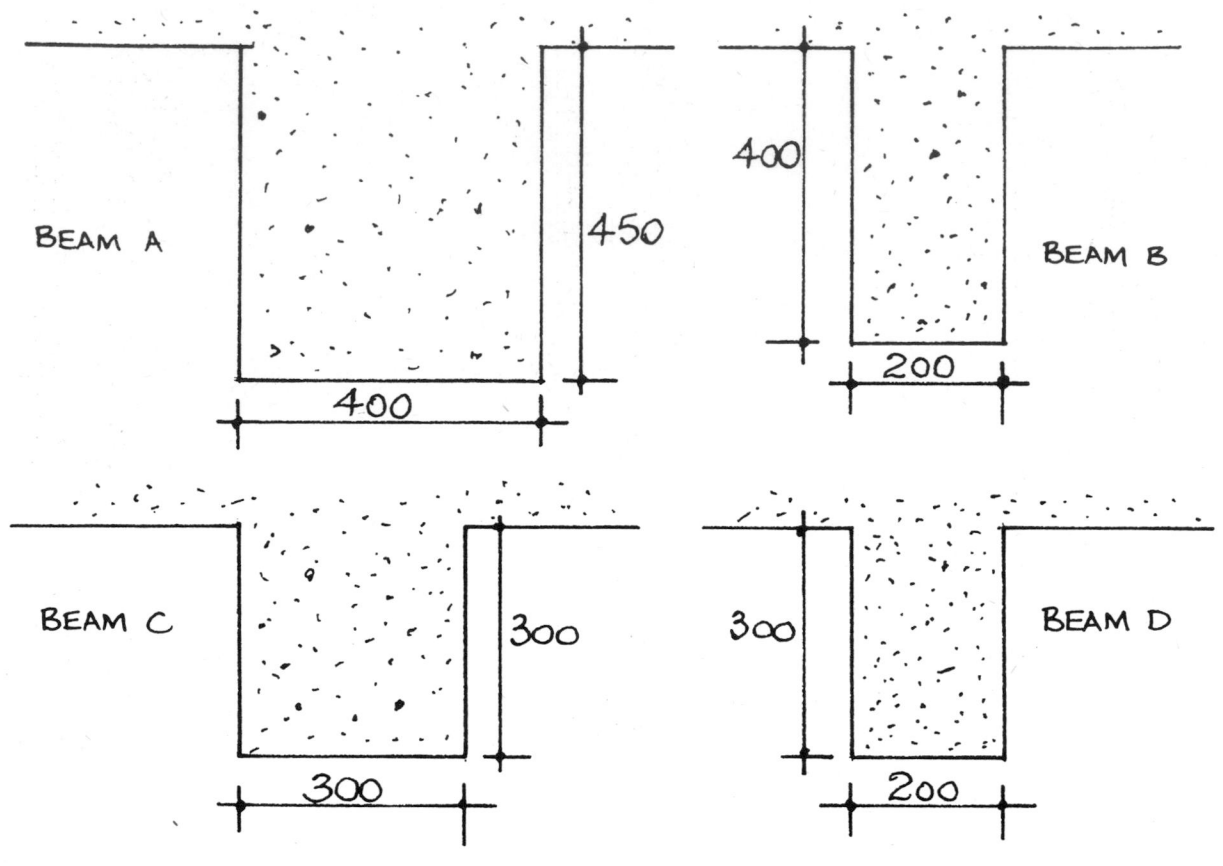

Assume each beam 10·000 long
Figure 1

Beam A

2/10·00	
0·45	
10·00	Two coat plaster ceiling as described
0·40	

Beam B

2/10·00	Two coat plaster ceiling as described
0·40	
10·00	Ditto not exceeding 300 mm wide
0·20	

Beam C

10·00	Two coat plaster ceiling as described not exceeding 300 mm wide
0·30	

Beam D

	Two coat plaster ceiling as described not exceeding 300 mm wide
2/10·00	
0·30	
10·00	
0·20	

2

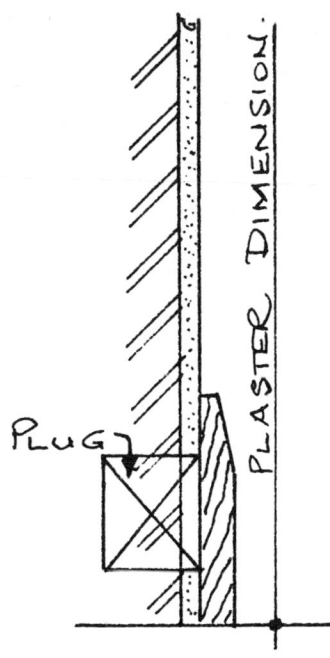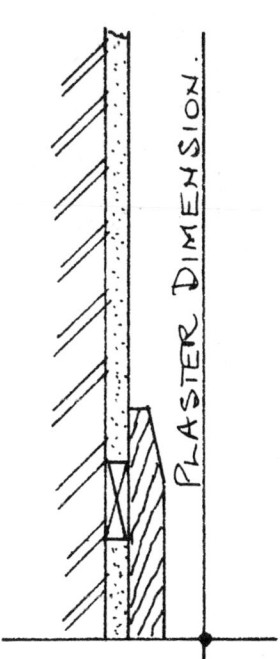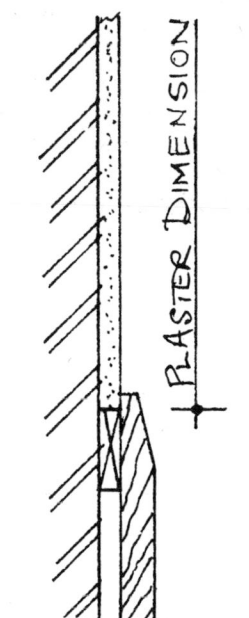

Figure 2

(6) Painting and decorating work is 'internal' or 'external' (SMM W1(a)) according to its position in the finished building (Practice Manual VI).

(7) The taker off normally marks each item of painting and decorating work as either internal or external but this may not be necessary where the group is obvious, e.g. wall papering. Alternatively he may include a sub-heading in order to avoid frequent repetition.

(8) Painting and decorating work shall be measured the area covered and allowances made for the extra girth of edges, mouldings, panels, sinkings, corrugations and the like (SMM V2). The taker off is the sole judge as to what comprises an appropriate allowance, which may be made by measuring the work as a flat surface and multiplying the dimensions by an appropriate factor.

(9) In addition to the usual protection items at the end of most trade sections the SMM requires individual items of protection for each type of flooring, floor finishing and paving, stating the area and type of work to be protected. No reference to any of these items is made during taking off as these are later incorporated by the worker up.

Plate 1
INTERNAL FINISHINGS 'A'

S.M.M. RULES

SECTION A
N.1.13.28.
T.3.4.5.8.13.14.
V.1.2.3.4.

APPROACH:-

Ceilings
Floors
Walls
Skirtings.

ALL NEW WORK INTERNAL.

(Ceiling.

6.50	
4.50	
2/4.50	
0.35	

Two ct Carlite plaster 10mm thk first ct in Bonding plaster 2nd ct in Finish plaster wi steel trowel finish on conc clg.

&

2ce emulsion plaster clg

S.M.M. T3. V1.
Use subheading to avoid repetition in descriptions.
Ceilings and floors measured independently since areas differ due to beams and piers.
S.M.M. T.3.4.5.
Measured area in contact with the base. S.M.M. T.4.   Description -
Material & composition   - Carlite Bonding & Finish plasters
Thickness   - 10mm
Coats   - Two
Surface treatment   - Steel trowelled
Base   - Concrete
Classification- Ceiling

S.M.M. V.1.2.3.4.

2/0.34	
0.11	

Ddt-
last two items

Area displaced by piers at boundary of measured area. S.M.M. A.3.

S.M.M. T.5.

	4.500
piers 2/0.113	0.226
	4.274

2/4.27

Extl angle plaster

External angles of beam.

(Floor

Total 0.040
P.V.C. 0.003
Bed 0.037

S.M.M. T.13.4.8.

6.50
4.50

Ct & sand (1:3) 37 mm thk level trowelled bed to receive p.v.c. tiles on conc

&

300 x 300 x 3 mm P.V.C. floor tiles laid wi symmetrical jts faced wi adhesive on level trowelled bed

Description –
Composition
& mix   – Cement & sand (1:3)
Thickness   – 37mm
Coats   – Contractor's discretion
Base   – Concrete
Classification– Level trowelled bed
                  to receive P.V.C. tiles.
S.M.M. T.8.   – Level.

S.M.M. T.14.   Description –
Kind   – P.V.C. tiles
Thickness
& size   – 300 x 300 x 3mm
Shape   – Rectangular
Bedding   – Adhesive
Joints   – Symmetrical
Base   – Trowelled bed S.M.M.T3.
Classification– Floor
S.M.M. T.8.   – Level.

2/0.34
0.11

Delt
last two items

Area displaced by piers at boundary
of measured area S.M.M. A.3.

(walls.

6.500
4.500
2/ 11.000
22.000

S.M.M. T.3.4.5.

Measured the area in contact with
the base S.M.M. T.4.

22.00
3.00

Two ct Carlite plaster 13 mm thk first ct in Perlited Browning plaster 2nd ct in Finish plaster wi steel trowel finish on bk walls

&

2ce emulsion plaster walls

S.M.M. V.1.2.3.4.

5

2/	0.34		$\underline{Ddt}$
	0.35		last two items

Area displaced by beam at boundary
of measured area. S.M.M. A.3.

$$\begin{array}{r} 3.000 \\ beam \quad 0.350 \\ \hline 2.650 \end{array}$$

S.M.M. T.3.4.5.

2/2/	2.65		Two ct plaster
	0.11		a.b. bk walls

n.e. 300 mm wide

Sides of attached piers regarded as
work to abutting walls.

&

2ce emulsion
plastr walls.

S.M.M. V.1.2.3.4.
Not isolated surface.
Work on piers added to similar work
on walls.

2/2/	2.65		Extl angle
			plastr

S.M.M. T.5.
External angles of attached piers.

(Sklg

S.M.M. N.13.28.

	22.00		25 × 100 mm S.W.
2/2/	0.11		chamf sklg pl
			to bkk

Particulars of quality given in
Preamble in preference to
description.

&

K.P.S. ③ sklg
n.e. 150 mm gth

S.M.M. V.1.2.3.4.
Particulars of materials given in
Preamble in preference to
description.

&

$\underline{Ddt}$
2ce emulsion
plastr walls

X 0.10 =

ie. not
exceeding
150mm girth

S.M.M. N.1.
Cross-sectional area exceeds
0.002 m²

	4		mitre sklg
2/2/	2		

6

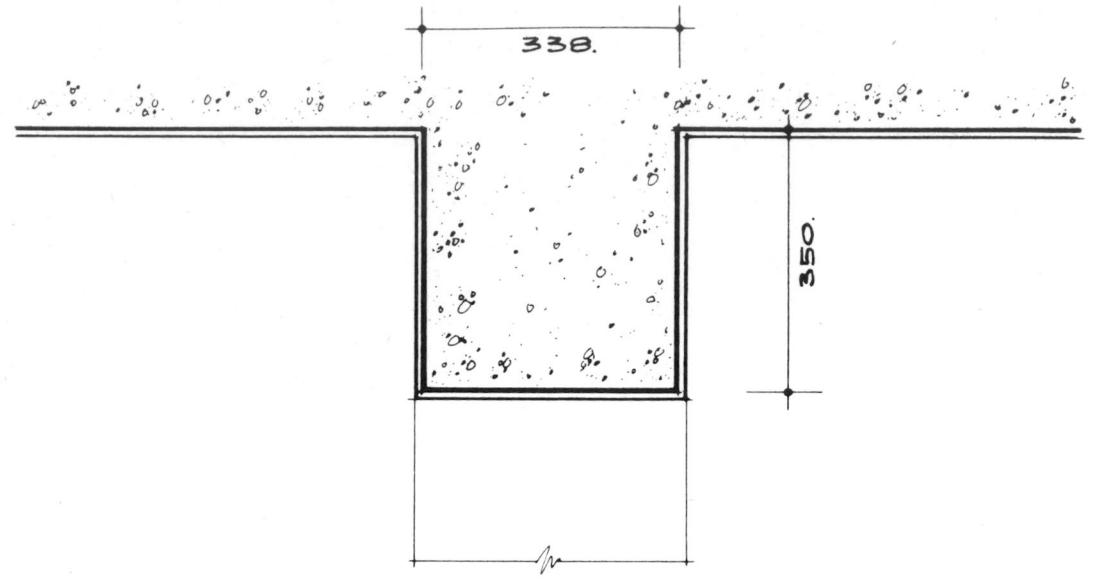

338.

350.

BEAM SECTION.

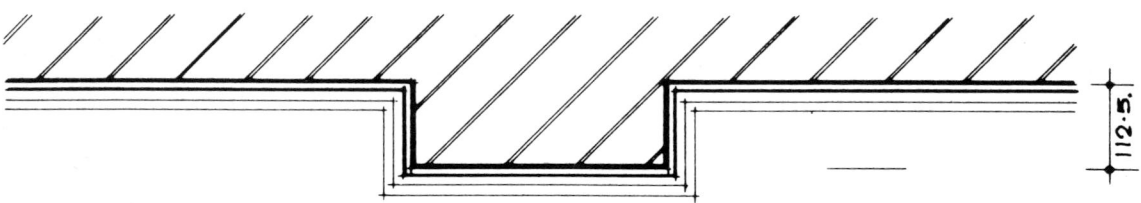

112·5.

PIER PLAN.

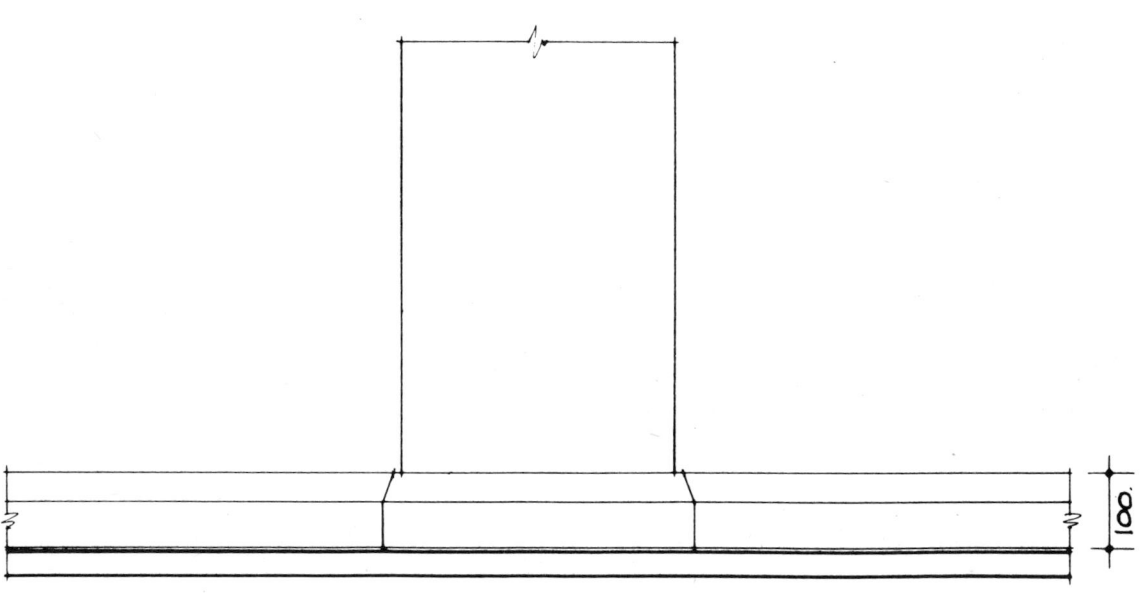

100.

PIER ELEVATION.

SCALE. 1 : 10.

Plate 1
INTERNAL FINISHINGS 'A'

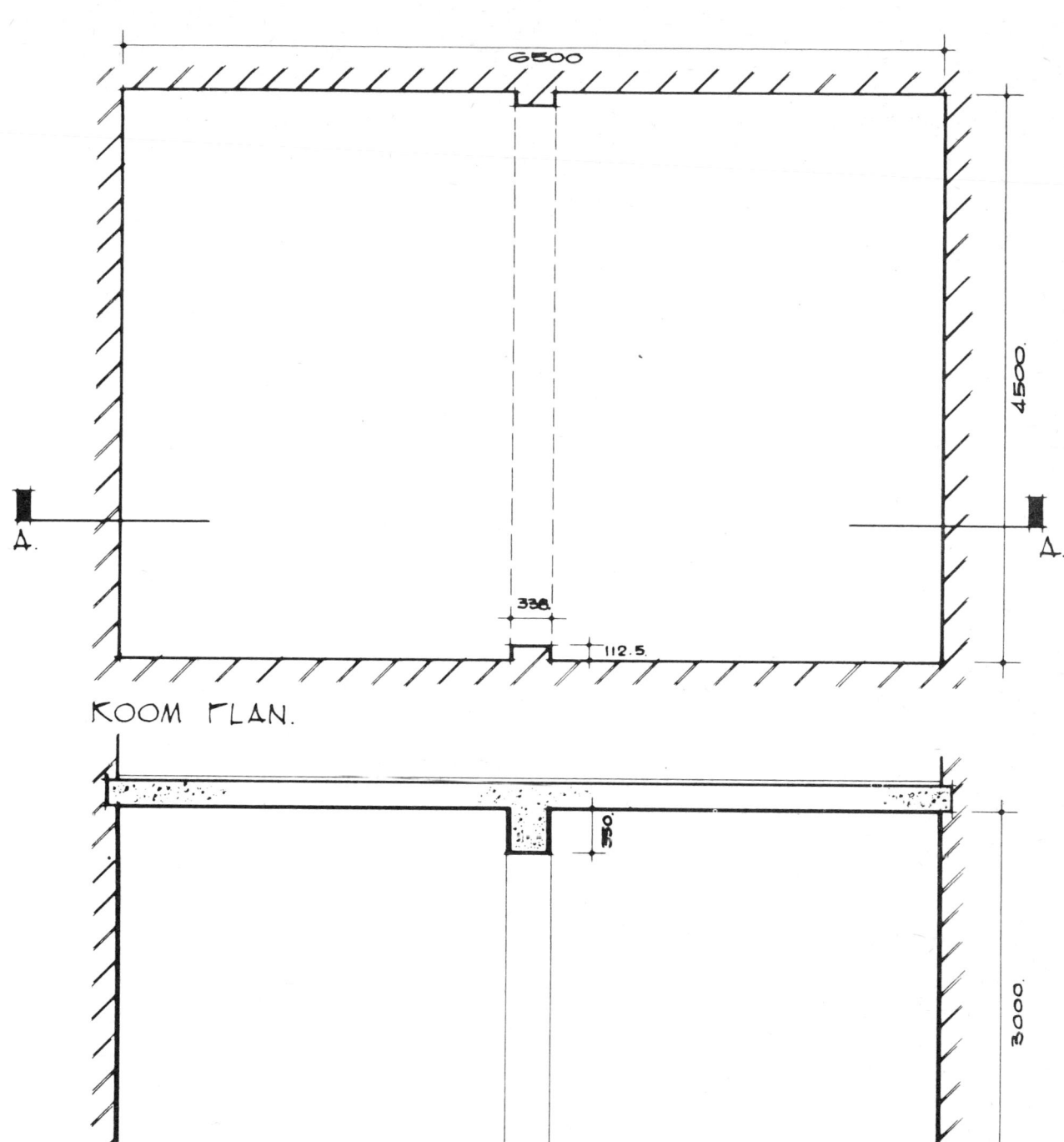

6500

4500.

A.   A.

336

112.5

KOOM FLAN.

350

3000.

SECTION A. A.

KEY FOK FINISHES.   SCALE. 1 :50.

CEILING.   Render in Carlite Bonding Plaster and Set in
                Carlite Finish Plaster 10mm total thickness
                Two coats Emulsion Paint.

FLOOK.   Total finish 40mm thick.
                300 x 300 x 3mm P.V.C. tiles on
                Cement and Sand (1 : 3) bed.

WALLS.   Render in Carlite Perlited Browning Plaster
                and Set in Carlite Finish Plaster 13mm total thickness,
                Two Coats Emulsion Paint.

SKIKTING.   25 x 100mm Softwood chamfered
                K.P.S. Two undercoats, one coat gloss.

Plate 2
INTERNAL FINISHINGS 'B'

S.M.M. RULES   SECTION A

             N.1.11.13.28.
             T.3.4.5.8.13.14.16.17.22.26.29.32.34.
             V.1.2.3.4.12.

APPROACH:-

             1. Lecture Hall   Measure each group of rooms
               Entrance Hall   in standard sequence:-
               Corridor.   Ceilings
             2. Studies.   Floors
             3. Store.   Walls
             4. Toilets   Skirtings.
               Kitchen.

         (Lecture Hall 01
         Entrance Hall 07
          Corridor   10.

ALL NEW WORK INTERNAL.

S.M.M. T3. V1.
Use subheading to avoid repetition
in descriptions.

        (Ceilings
        & floors

Ceilings and floors measured
together since areas are identical.

      4/2.400 = 10.800
      3/0.113 =   0.339
                 11.139

S.M.M. T.4.
Measured area in contact with
the base.

S.M.M. T.16.17.   Description-

Kind & quality	– gypsum plasterboard
Thickness	– 9.5mm
Base	– softwood
Fixing	– galv. clout nails
Joints	– hessian scrim cloth.

13.50
5.50
3.40
3.50
11.14
1.50

9.5mm thick
gypsum plasterboard
clg fxd to s.w. wi
galv clout nails
the jts covered wi
hessian scrim cloth

&
One ct gypsum
plaster 3mm thk
wi steel trowel
fin on plasterbd
clg

S.M.M. T3.4.5.

Material & Composition	– gypsum plaster
Thickness	– 3mm
Coats	– one
Surface treatment	– steel trowelled
Base	– plasterboard
Classification	– ceiling.

Item			

Prime Cost Sum

£ _____

for Embossed
clg paper

&

Add Profit

S.M.M. V12. A8.

WORKER UP to insert the total cost
as calculated from the net area of
hanging ceiling paper.

13.50	
5.50	
3.70	
3.50	
11.14	
1.50	

Size plaster clg
& hang embossed
paper (In No _____
pieces £1.50 per
piece).

S.M.M. V12.
The number of pieces to be
calculated from the net area of
hanging ceiling paper.

&

2ce emulsion ditto.

Total 0·05
Blocks 0·025
Bead 0·025

S.M.M. V.1.2.3.4.

&

Ct & sand (1:3)
25 mm thk level
floated bed to
receive wood block
floor on conc

S.M.M. T.13.4.8.

&

225 x 75 x 25 mm
Sapele t & g. block
floor laid basket
pattern wi two
block plain border
in whole units
in Synthaprufe
adhesive on level
floated bed,
surface fin wi
two coats wax
polish

S.M.M. T14.   Description-
Kind   - sapele t.&g. blocks
Size &
thickness   - 225 x 75 x 25mm
Shape   - rectangular
Finish   - wax polish
Bedding   - adhesive
Joints   - basket pattern with
                    plain border.
Base   - floated bed (SMM T3)
Classification- floor
S.M.M. T8.   - level.

Polishing included in description
NOT measured separately as SMM V4.
(Practice Manual V4).

		(Walls
		13.500
		5.500
	2/	19.000
		38.000
		3.700
		3.500
	2/	7.200
		14.400
		11.139
		1.500
	2/	12.639
		25.248

<table>
<tr><td>38.00
3.30</td><td rowspan="3">⌉
⌋</td><td>Two coats gypsum plaster 15mm thk. first ct in Browning plaster 2nd ct in Finish plaster w/ steel trowel fin on bk walls</td><td>S.M.M. T3.4.5.
Measured area in contact with the base SMM T4.</td></tr>
<tr><td>14.40
2.50</td></tr>
<tr><td>25.28
2.50</td></tr>
</table>

38.00
3.30

14.40
2.50

25.28
2.50

⌉
│
│
│
⌋

Two coats gypsum
plaster 15mm thk.
first ct in Browning
plaster 2nd ct in
Finish plaster w/
steel trowel fin
on bk walls

&

Size plaster wall
& hang pattern
wall paper (In
No. pieces £1.00
per piece).

S.M.M. V12.
The number of pieces to be
calculated from the net area of
hanging wall paper.

Item

Prime Cost Sum

£ _____

for pattern wall
paper

&

Add Profit

S.M.M. V12. A8.
WORKER UP to insert the total cost
as calculated from the net area of
hanging wall paper.

S.M.M. T3.4.5.
Measured area in contact with
the base SMM T4.

12

38.00	Plasterboard Cove	S.M.M. T22.26.	
14.40	125 mm gth butt		
25.28	jtd & fixed wi		
	plaster adhesive		

&

2ce emulsion
plaster clg

X 0.13 =

S.M.M. V1.4.
Cove painted in conjunction with
ceiling and not work on isolated
surfaces SMM V3.

3/ 4   Angles cove   S.M.M. T26.

	38.000
4/0.100	0.400
	37.600
	14.400
	0.400
	14.000
	25.278
	0.400
	24.878

37.60	Ddt	Area displaced by cove.
0.10	Hang embossed	
14.00	clg paper a.b.	
0.10		
24.88	&	
0.10		

Ddt
2ce emulsion ditto

38.00	Ddt	Area displaced by cove.
0.10	Hang pattern	
14.40	wall paper a.b.	
0.10		
25.28		
0.10		

(Sktgs

38.00			
14.40			
25.28			

25 × 100 mm Sapele chamf. sktg selected & kept clean for polishing scrd. csk & pellated to sw.

&

15 × 50 mm sawn sw. grnd pl to bkk.

&

S.M.M. N1.11.13.28.
Particulars of quality given in Preambles in preference to description.

Two cts wax polish wood sktg n.e. 150 mm gth

&

S.M.M. V1.3.

Ddt:
Two cts plaster walls a.b.
× 0.10 =

&

Ddt
Pattern wall paper ab
× 0.10 =

3/ 4   hitoes sktg.

S.M.M. N1.

Studies
02 — 05

(Ceilings
& floors

14

4/	3·40			
	2·40			

9.5 mm Thk
Gypsum plasterbd
clg a.b.

&

One ct gypsum
plaster 3 mm thk
a.b.

&

2ce emulsion ditto.

Total 0.050
Finish 0.020
Base 0.030

&

Ct & sand (1:3)
30 mm thk level
trowelled bed to
receive carpet
on conc.

&

Oliver plain
cut pile Wilton
woadloom floor
carpet (messrs Firth
Carpets Ltd) fxd wi
tackless grippers
& heat bonded
jts on latex
underlay

```
        3·400
        2·400
    2/ 6·4·00
       12·800
```

4/ 12·80

Perimeter fxg wi
tackless grippers
pl to trowelled
bed

S.M.M. T16.

S.M.M. T3.4.5.

S.M.M. V1.2.3.4.

S.M.M. T13.4.8.

S.M.M. 29.32.
Wall to wall dimensions.
Description-
Kind — Broadloom
Quality — Oliver cut pile Wilton
(Firth Carpets Ltd.)
Pattern — plain
Fixing — tackless grippers
Joints — heat bonded if reqd.
Base — latex underlay.

S.M.M. T32.

		grippers 2/0.040	3.700	S.M.M. T34.
			0.080	Measure to inside face of grippers.
			3.620	
			2.700	
			0.080	
			2.620	

CARPET
UNDERLAY

4/ 3.62
2.62

"Duralay" latex underlay laid loose on level trowelled bed

```
        3.700
        2.700
    2/  6.400  (Walls
       12.800
```

4/ 12.80
2.50

Two ct plaster walls a.b.

&

2ce emulsion ditto

S.M.M. T3.4.5.

S.M.M. V1.3.4.

(Sktgs

4/ 12.80

25 x 100 mm s.w. chamf sktg pl to bkk

&

K.P.S. ③ wood sktg n.e. 150 mm gth

&

Ddt

2ce emulsion plaster walls

X 0.10

S.M.M. N13.28.
Particulars of quality given in Preambles in preference to description

S.M.M. V1.2.3.4.
Particulars of materials given in Preambles in preference to description.

4/ 4

mitres sktg

S.M.M. N1.

		(Store 06			
		(Clg &			
		(floor			

3.70
2.10
―――

9.5 mm Thk
gypsum plastered
clg a.b.

S.M.M.T16.

&

One ct. gypsum
plaster 9mm thk
ab

S.M.M. T3.4.5.

&

2ce emulsion ditto

S.M.M. V1.2.3.4.

&

Granolithic conc.
50 mm thk floor
wi steel trowel
finish on level
conc bed

S.M.M. T4.8.3.
Particulars of composition and mix
given in Preamble in preference to
description.

(Walls

	3.700
	2.100
2/	5.800
	11.600

11.60
2.50
―――

Two coat plaster
walls a.b.

S.M.M. T3.4.5.

&

2ce emulsion ditto

S.M.M. V1.3.4.

(Sktg

11.60
―――

25 x 100 mm S.W.
chamf sktg a.b.

S.M.M. N13.

&

K.P.S. ③ wood
sktg ab

S.M.M. V1.2.3.4.

&

Ded:
2ce emulsion plaster
walls x 0.10 = ―――

	4	mitres sktg		S.M.M. N1.	
		(Toilet 08			
		(Kitchen 09			
		(Clg &			
		(floor			
	2/3.40	9.5 mm Gypsum		S.M.M. T16.	
	2.60	plasterbd clg ab			
		&			
		One ct. gypsum		S.M.M. T3.4.5.	
		plaster 3mm thk ab			
		&			
		2ce eggshell		S.M.M. V1.2.3.4.	
		finish ditto			
		&			
		0.050			
		tiles 0.013			
		bed 0.010 0.023			
		bed 0.027			
		Ct + sand (1:3)		S.M.M. T13.4.8.	
		27 mm thk level			
		screeded bed to			
		receive clay tiles			
		on conc			
		&			
		150 × 150 × 13 mm Red		S.M.M. T14.	
		plain clay floor			
		tiles laid wi			
		symmetrical jts			
		bedded in ct +			
		sand (1:3) on level			
		screeded bed			
		the jts grouted			
		in neat ct.			

(Walls

	3.700
	2.600
2/	6.300
	12.600

S.M.M. T13.4.5.
Measured the area in contact with
the base (SMM T4).

2/ 12.60
2.50

Ct + sand (1:3)
15 mm thk
floated backing
to receive ceramic
wall tiles on bkk.

walls	12.600	
backg	0.015	
tiles	0.005	
4/2/	0.020 =	0.160
		12.440

S.M.M. T14.
Measured on the exposed face
(SMM T4).

2/ 12.44
2.50

100 × 100 × 5 mm
cream glazed
ceramic wall tiles
laid w/ symmetrical
jts fxd w/
adhesive to
floated backing
the jts grouted
in white ct.

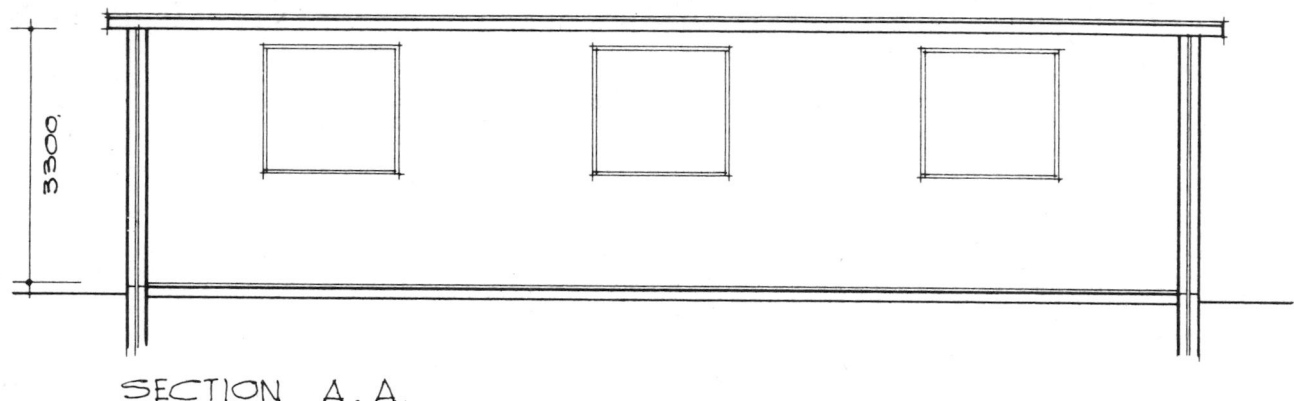

SECTION A.A.

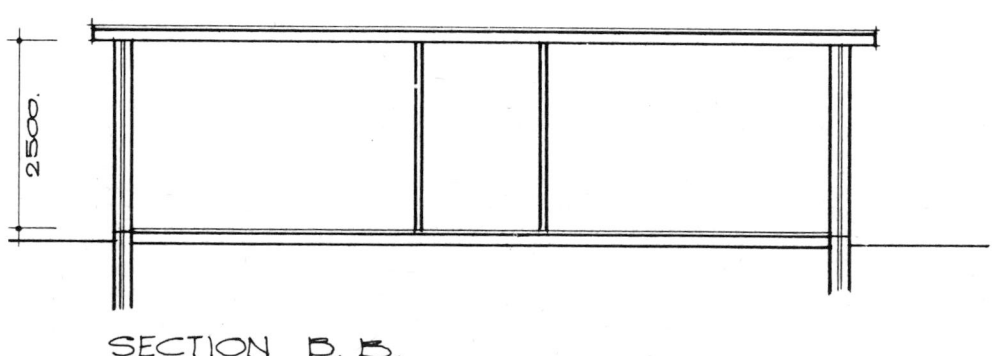

SECTION B.B.

Plate 2
INTERNAL FINISHINGS 'B'

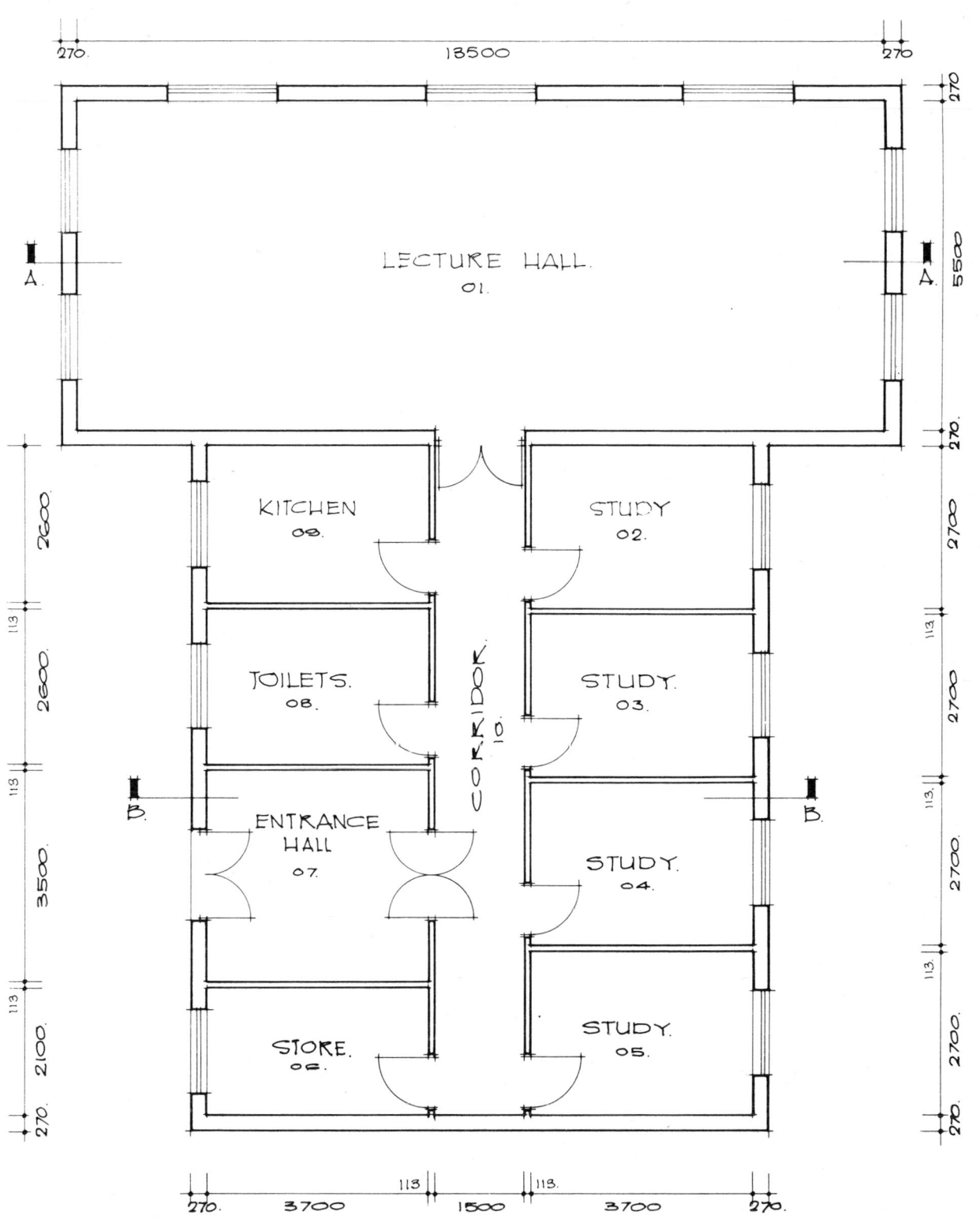

270   18500   270

270

A   LECTURE HALL   A   5500
       01.

270

2600   KITCHEN   STUDY   2700
        09.   02.

113   113

2600   TOILETS.   C   STUDY.   2700
        08.   O R R I D O R   03.
              10.

113   113

B   ENTRANCE   STUDY.   B
       HALL   04.   2700
3500   07.

113   113

2100   STORE.   STUDY.   2700
        06.   05.

270   270

270   3700   113   1500   113   3700   270

SCALE. 1 :100.

KEY FOR FINISHES		
CEILINGS		
	C1	9·5 mm Gypsum plasterboard and skim coat in neat gypsum plaster. Embossed ceiling paper P.C. £1·50 per piece. Two coats emulsion paint.
	C2	9·5 mm Gypsum plasterboard and skim coat in neat gypsum plaster. Two coats emulsion paint.
	C3	9·5 mm Gypsum plasterboard and skim coat in neat gypsum plaster. Two coats eggshell finish.
	C4	Plasterboard cove 125 mm girth. Two coats emulsion paint.
FLOORS		
		Total finish 50 mm thick.
	F1	225 X 75 X 25 mm Sapele block floor. Cement and sand (1:3) bed.
	F2	Cut-pile Wilton broadloom Oliver plain carpet (Messrs Firth Carpets Ltd.,) with "Duralay" latex underlay (20 mm over-all thickness). Cement and sand (1:3) bed.
	F3	50 mm Thick granolithic concrete.
	F4	150 X 150 X 13 mm Red clay floor tiles bedded in 10 mm cement and sand. Cement and sand (1:3) bed.
WALLS		
	W1	Render and set gypsum plaster. Two coats emulsion paint.
	W2	Render and set gypsum plaster. Patterned wall paper P.C. £1·00 per piece.
	W3	100 X 100 X 5 mm Coloured ceramic wall tiles fixed with adhesive on 15 mm cement and sand (1:3) backing.
SKIRTINGS		
	S1	25 X 100 mm Softwood chamfered. K.P.S. two under-coats, one coat gloss.
	S2	25 X 100 mm Sapele chamfered. Two coats wax polish.

FINISHES: SCHEDULE KEY

ROOMS		CEILINGS	FLOORS	WALLS				SKIRTING	
No.	DESCRIPTION			N	S	E	W		
01	Lecture Hall	C1 C4	F1	W2	W2	W2	W2	S2	
02	Study	C2	F2	W1	W1	W1	W1	S1	
03	Study	C2	F2	W1	W1	W1	W1	S1	
04	Study	C2	F2	W1	W1	W1	W1	S1	
05	Study	C2	F2	W1	W1	W1	W1	S1	
06	Store	C2	F3	W1	W1	W1	W1	S1	
07	Entrance Hall	C1 C4	F1	W2	W2	W2	W2	S2	
08	Toilets	C3	F4	W3	W3	W3	W3	–	
09	Kitchen	C3	F4	W3	W3	W3	W3	–	
10	Corridor	C1 C4	F1	W2	W2	W2	W2	S2	

FINISHES: SCHEDULE

Plate 3
EXTERNAL FINISHINGS

S.M.M. RULES	SECTION A M.3.4.5.6.8.17.43.49.50. N.1.3.4. V.1.2.3.4.	
APPROACH:-	Rendering and labours. Tilehanging & labours/ boarding & labours.	
	———————	
6·30 <u>0·68</u>	Two ct external rendering 20mm thk first ct in mortar (1:1:6) 2nd ct in "Cullamix" Tyrolean finish on conc block walls	S.M.M. T3.4.5. Classified as external work (SMM T3).
6·30 <u>1·13</u>	Vertical coverings of 265 x 165 mm machine made "CUB" pattern clay tiles to 40mm lap each tile nailed every course wi two aluminium alloy nails on & incl 38 x 19 mm tanalised sw. battens	S.M.M. M3.4.5.6. Description- Kind of tile – clay Size – 265 x 165 Type – club pattern Quality – machine made lap – 40mm Fixing – two nails each tile Battens – 38 x 19mm softwood Classification– vertical coverings.
	&	– Preliminary treatment of timber S.M.M. N1.
	Underlay of reinf bituminous felt lapped 150 mm at jts nailed to sw wi galv. clout nails	S.M.M. M17.

| | 6·30 | | E.O. double cos wi plain eaves tile | | | S.M.M. M8. Bottom edge vertical tiling. |
| | | | ↯ | | | |

E.O. double
cos wi plain
eaves tile

↯

milled lead
Code No 4 BS 1178
flashing 200 mm
gth wi edge
shaped to match
"CurB" tiles wi
lead clips & tacks
& 100 mm
intermediate laps
Copper nailed

↯

Copper nailing
@ 50 mm c/c

S.M.M. M8.
Bottom edge vertical tiling.

S.M.M. M48.49.50.
Top edge vertical tiling.

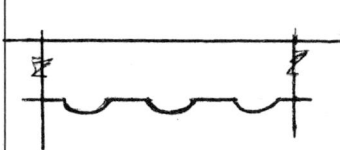

S.M.M. M43.

25

6·30	Western Red Cedar weatherboarding 25 mm extreme thickness in 150 mm widths wrot on one face and one edge, rebated and lapped 25 mm at jts. wi splayed heading jts, secret nailed wi galv nails to vertical walls external & clean off on completion	S.M.M. N1.4.5. Description- Kind & Quality- Western Red Cedar weatherboarding
1·13		Surface treatment - Wrot one face, one edge.
		Thickness - 25mm extreme
		Jointing - rebated & lapped 25mm
		Fixing - secret nailed
		Classification- vertical walls external.
	&	
	Underlay ab	S.M.M. M17.
1 10/ 6·30	One ct "Cuprinol" Cedar board walls	S.M.M. V1.2.3.4.
1·13	new work E×12	Note allowance for extra girth of edges (SMM V2).

26

Plate 3
EXTERNAL FINISHINGS

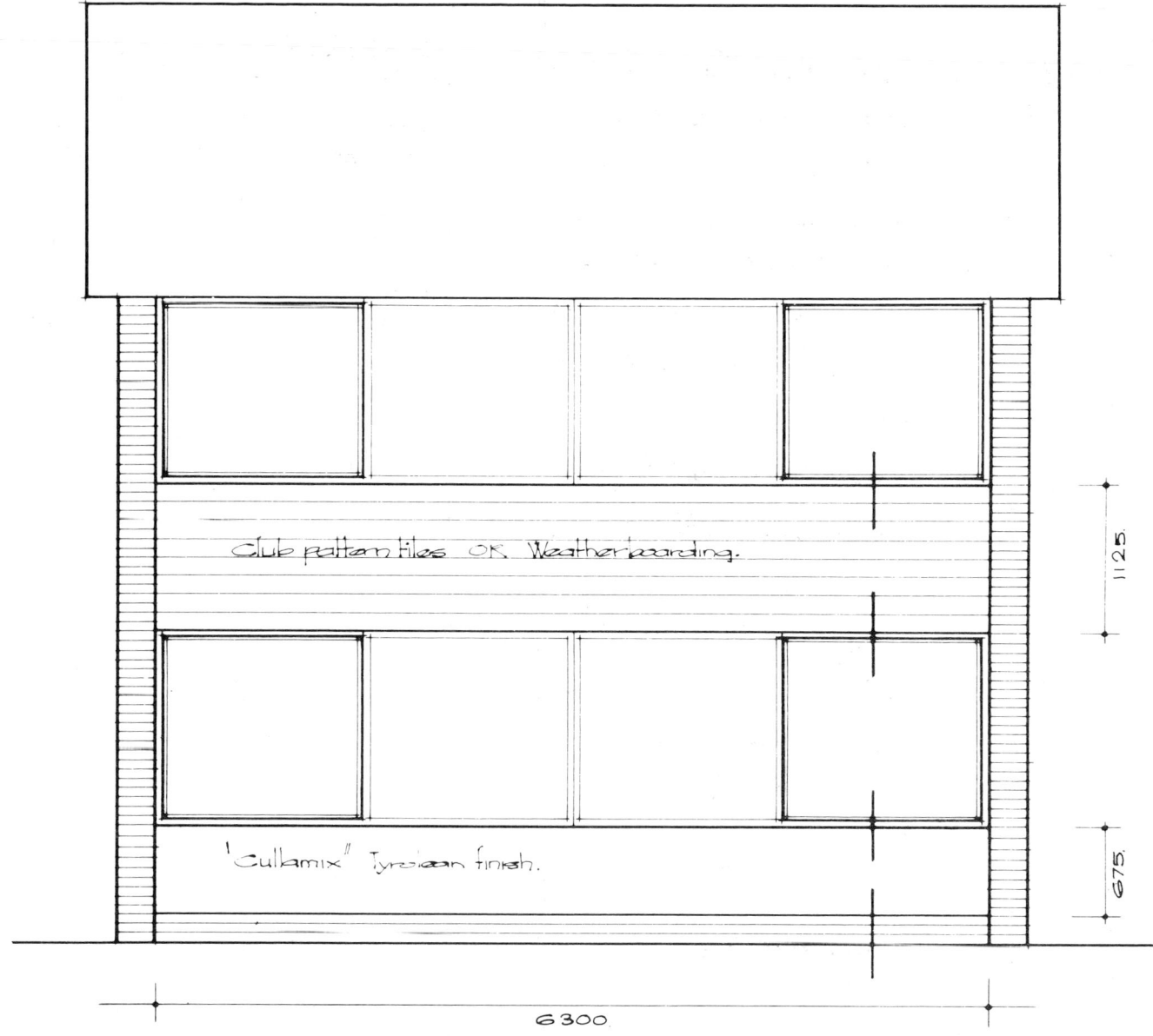

Club pattern tiles OR Weatherboarding.

'Cullamix' Tyrolean finish.

1125

675.

6300

SCALE . 1 : 50.

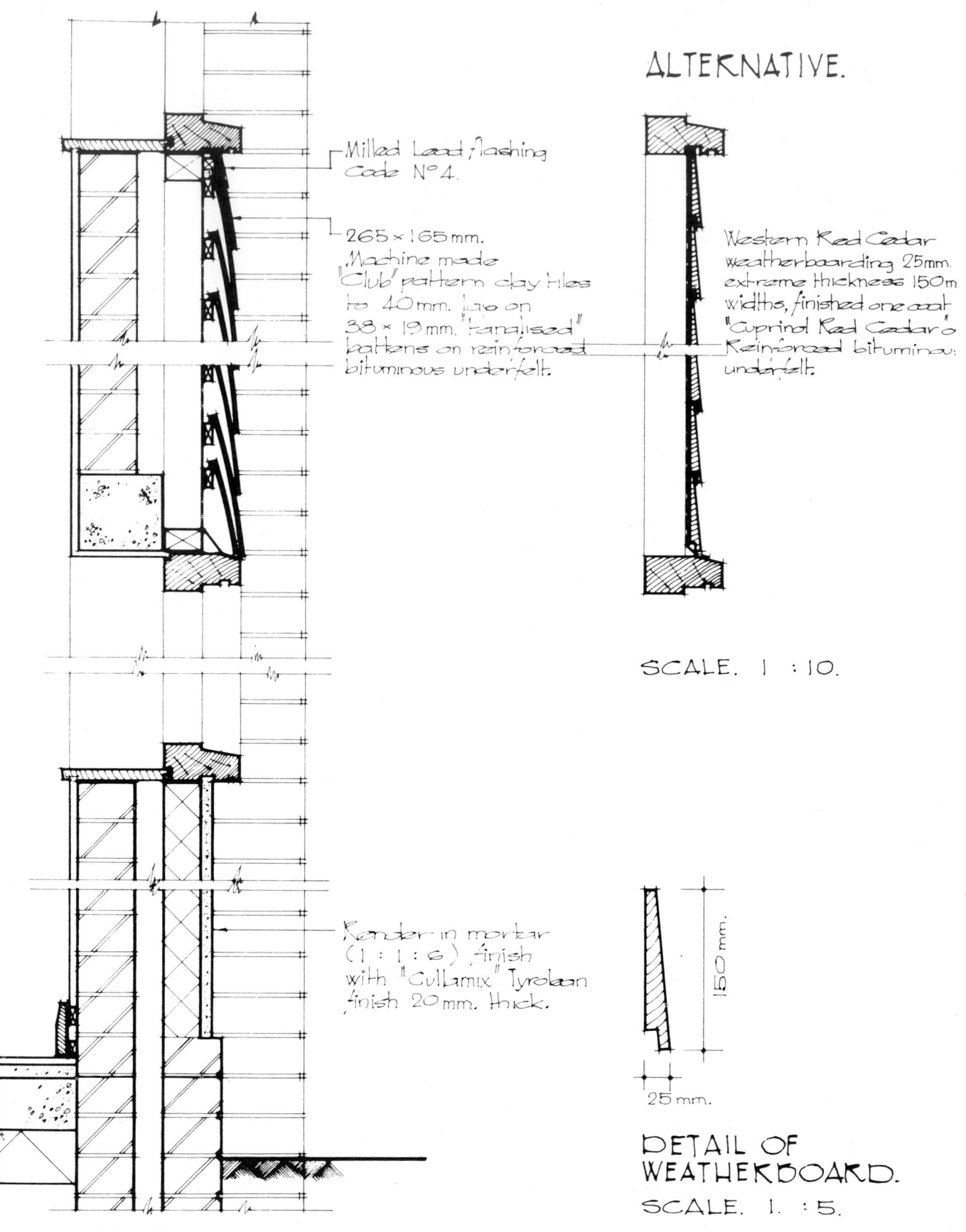

ALTERNATIVE.

Milled Lead flashing Code Nº 4.

265 × 165 mm. Machine made "Club" pattern clay tiles to 40 mm. lap on 38 × 19 mm. "tanalised" battens on reinforced bituminous underfelt.

Western Red Cedar weatherboarding 25mm. extreme thickness 150m widths, finished one coat "Cuprinol Red Cedar". Reinforced bituminous underfelt.

SCALE. 1 : 10.

Render in mortar (1 : 1 : 6) finish with "Cullamix" Tyrolean finish 20 mm. thick.

150 mm.

25 mm.

DETAIL OF WEATHERBOARD.

SCALE. 1 : 5.

DETAIL AT A.A.

SCALE. 1 : 10.

Chapter II

Windows

Approach

(1)   Where a large project includes several buildings measure each as a separate unit.

(2)   Compare the window schedule with the drawings to ensure that all windows have been given a reference number and are included in the schedule. Schedules provide the essential information in a concise, tabulated form for easy reference, as a more practical alternative to embodying the details throughout the drawings. This saves time involved in searching for information, and where schedules are not provided by the architect the taker off is well advised to prepare his own. The preferred forms of schedule recommended by BS 1192 'Building Drawing Practice' are illustrated in Appendix C. Many surveyors prefer to prepare their own schedules giving particulars of windows and openings to enable them to measure direct from the schedules without further reference to the drawings.

The use of a schedule avoids unnecessary repetition when taking off, since windows with similar features may be readily identified and grouped together.

(3)   The measurement of windows normally includes the adjustment of walls and finishings which have been measured over the window openings.

(4)   Each window must be considered on its merits but a typical sequence of measurement would be:

Window	casement (or sash)
	ironmongery
	glass
	fixing window
	painting
Opening adjustment	walls and finishings
	head inside and out
	reveals inside and out
	sill inside and out

Measurement of windows

METAL WINDOWS. Standard metal windows shall each be enumerated separately (SMM Q2) and are normally identified by a reference number relating to the appropiate BS (SMM A6).

Special metal windows (complete with frames, mullions, transomes, hinges and fastenings) shall each be enumerated separately stating the overall size, the nature of the construction, the shape where other than rectangular, the finish and the number of opening portions (SMM Q3).

Building in metal windows (complete with frames) shall be enumerated stating the overall size. Building in or cutting and pinning lugs, bedding frames and pointing to one or both sides shall be given in the description. (SMM G48).

WOOD WINDOWS. Stock-pattern items (e.g. casement windows, sash windows) shall be enumerated stating the appropriate reference (SMM A6). The item is described in the form in which it is likely to be delivered to the work and unless otherwise stated incorporation into the works shall be deemed to be included. Glazing beads, glazing fillets and ironmongery supplied with stock pattern items shall be included in the description (SMM N17). Additional ironmogery must be measured separately in accordance with SMM N32. Stock-pattern items are normally identified by a reference number relating to the manufacturers catalogue.

Composite items, other than stock pattern items, are each enumerated separately in the form they are likely to be delivered to the works (SMM N17) and unless they can be adequately described must be accompanied by a bill diagram (SMM N21). An adequate description of al the component parts would inevitably produce a very long description and the taker off is therefore advised to include a bill diagram. As with stock pattern items, any ironmongery supplied with the window is included in the composite item and any other ironmongery is measured separately in accordance with SMM N32.

FIXING WINDOW. The measurement of glass is followed by the items for fixing the window including cramps or fixing slips, bedding and pointing, cover fillets and the like.

PAINTING. Painting and decorating work is 'internal' or 'external' according to its position in the finished building (Practice Manual VI). Work carried out on members before they are fixed shall be so described stating if carried out on or off site (SMM V1).
Painting and decorating work shall be measured the area covered and appropriate allowances made for the extra girth of edges, mouldings, panels, sinkings, corrugations and the like (SMM V2). The taker off is the sole judge as to what comprises an appropriate allowance, which may be made by measuring the work as a flat surface and multiplying the dimensions by an appropriate factor.

Work on windows (measured on each side over frames, linings, mullions, transomes sills and glass) is given in square metres stating the size of the panes as SMM V5.

Work on edges of opening casements shall be given in metres irrespective of girth. Additional painting (caused by opening lights) on the surrounding frame shall be deemed to be included with the item (SMM V5).

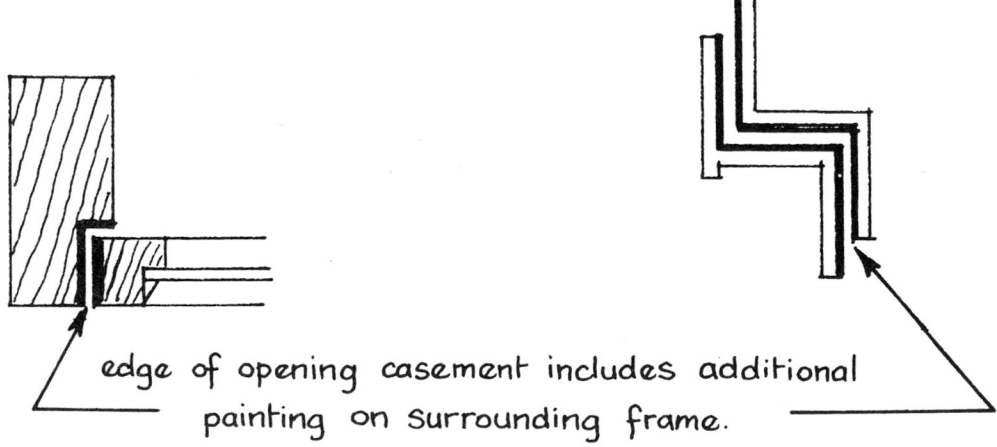

edge of opening casement includes additional painting on surrounding frame.

Figure 3

OPENING ADJUSTMENT. Having completed the measurement of windows, the taker off knows the exact size of the window openings and can then deduct the walls and finishings which have been measured over the opening. When measuring very small windows he should bear in mind that finishings and decoration are not deducted in the case of voids not exceeding 0·50 m² (SMM Sections T and V).

Deductions of brickwork, facework or blockwork for lintels shall be measured as regards height to the extent only of the full courses displaced (SMM G5, 14, 26). This is particularly important in the case of blockwork where the height of the lintel is frequently less than the height of a full course, so that no deduction is made. Furthermore no deduction is made for brickwork or blockwork displaced by the ends of lintels built into the wall (SMM G48).

Openings in cavity walls require a cavity gutter at the window head (SMM G37) and an item for closing the cavity at the jambs (SMM G9) although these are seldom indicated on the drawing.

Plate 4
STANDARD STEEL WINDOW

S.M.M. RULES	SECTION A	
	F.18.21.	
	G.3.5.14.48.	
	Q.2.	
	T.10.14.	
	U.2.3.4.	
	V.1.2.3.4.5.	

APPROACH:—

Window	— Standard window	
	Glass	
	Fixing	
	Painting	
Opening		
adjustment	— Walls & finishings	
	Head inside & out	
	Reveals inside & out	
	Cill inside & out.	

_____

(Mdw

1	Standard steel window B.S. 990 Type 9FV9	S.M.M. Q2. A6. Standard windows are supplied as complete units with opening casements already hung and fitted with ironmongery. Fixing lugs and screws are provided by the manufacturer and sent loose.

$$0.821 \times 0.143 = 0.114 \, m^2$$

0.82	3mm Thk clear sheet glass & glazg to metal wi metallic putty & metal clips in panes 0.10 — 0.50 m²	S.M.M. U2.3.4. Kind and nominal thickness — 3mm thick clear sheet
0.14		Compound — metallic putty
		Glazing — putty
		Securing — clips
		Sash — metal
		Classification— 0.10 — 0.50m²

$$0.848 \times 0.666 = 0.565 \, m^2$$

Classification— 0.50 — 1.00m²

0.85	Ditto in panes 0.50 — 1.00 m²	
0.64		

1	Build in metal wdw o/a size 900 × 900 mm incl bi. lugs beddg frame + ptg one side in mastic	S.M.M. G48.

Glass 0·114 m²
     0·565 m²
  2) 0·682
  Avg = 0·341 m²
  = MEDIUM PANES

0·90
0·90

Prime ② metal
wdw in medium
panes (new wk)
(INTL)

&

Ditto (new wk)
(EXTL)

Glass 0·821
      0·143
   2/ 0·964
      1·928

1·93

Ditto
edges of opg
casements (new wk)
(EXTL)

Opg
Adjustment

0·90
0·90

Ddt
One bk wall fcgs
one side fair face
other side ab

&

Ddt
2ce emulsion ditto
(INTL)

wdw 0·900
B.I. 2/0·100 0·200
      L = 1·100

1

Precast conc
(1:2:4) lintol size
1100 × 225 × 65 mm
finished fair 355 mm
gth reinf wi 2No
12 mm ∅ m.s. bars
& bi. to bkk in
mor (1:1:6)

— S.M.M. V1.2.3.5.

Pane size averaged as SMM V5.
Particulars of materials given in
Preamble in preference to
description.
This item includes painting on
ironmongery (SMM V5).

S.M.M. V5.
Use glass dimensions

S.M.M. G14.
Deduct overall size of window

S.M.M. V1.4.

S.M.M. F18.21.
Particulars of materials given in
Preamble in preference to
description.

65	225	65

Fair face = 355 mm
                  girth

Description quite adequate-
bill diagram not required.

S.M.M. F3.
Classification- other concrete work.

33

0·90 0·07	Ddt One bk wall fегs one side fair face other side ab	S.M.M. G5.14.48. No deduction made for ends of lintol.
2/0·90	Fair return fair face c.b. H.B. wide	S.M.M. G14. Internal reveals
3/0·90 0·11	2ce emulsion bk walls a.b.	S.M.M. V3.4. Not isolated surface
2/0·90	Fair return fcg bks H.B. wide	S.M.M. G14. External reveals
0·90	Red plain clay tile wdw cill 120 mm wide 13mm thk wi sndd edge bedd on bkk & ptd in mor (1:1:6)	S.M.M. T10.14.
2	Ends	S.M.M. T10.
1	$\begin{array}{r}\text{cfg } 0.900\\ \text{BI } 2/0.100 \quad 0.200\\ \hline L = \quad 1.100\end{array}$ Precast conc (1:2:4) sunk weath & throated cill size 1100 × 150 × 65 mm finished fair 210mm gth wi 2 No stooled ends & b.i. bkk in mor (1:1:6)	S.M.M. F18.21. Particulars of materials given in Preamble in preference to description. Description quite adequate – bill diagram not required. S.M.M. F3. Classification – other concrete work.

34

0.90 0.04			_Ddt·_ One lik wall fcgs one side fair face other side a.b. $\oint$		S.M.M. G5.14.	
			Add H.B. wall in c.b. stretcher bond in mor (1:1:6) ab $\oint$		S.M.M. G3.5. Behind concrete cill.	
			Add E.O. cb for fair face & flush ptg.		S.M.M. G14.	

35

FACING BRICKWORK.

FAIR FACE COMMON
BRICKWORK FLUSH POINTED
AND TWO COATS EMULSION.

LINTEL 2 N° 12mm. BARS.

13mm. THICK PLAIN
CLAY TILE CILL
120 mm WIDE.

SECTION A.A.

SCALE. 1:5.

Plate 4
STANDARD STEEL WINDOW

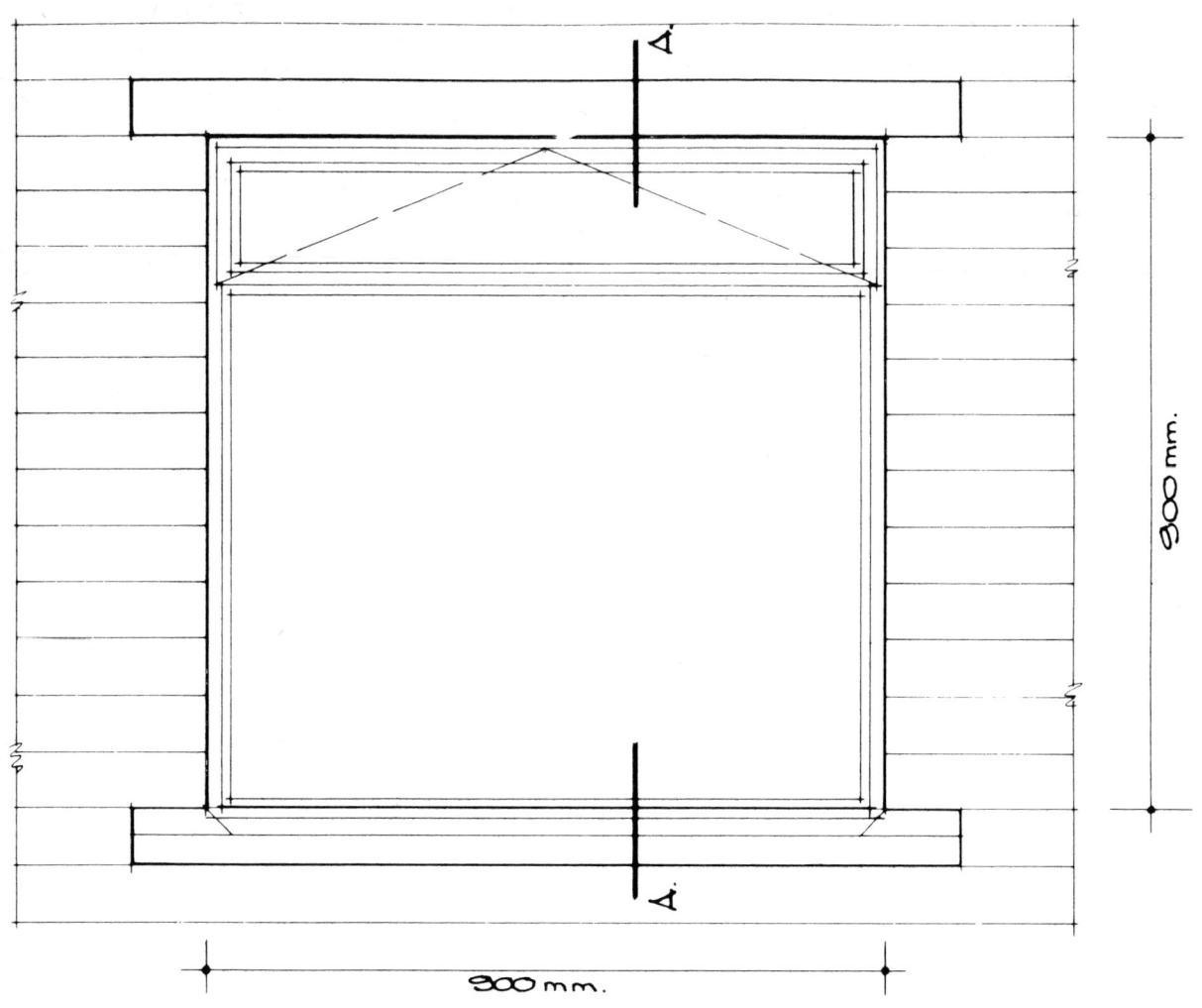

ELEVATION.   SCALE 1 : 10.

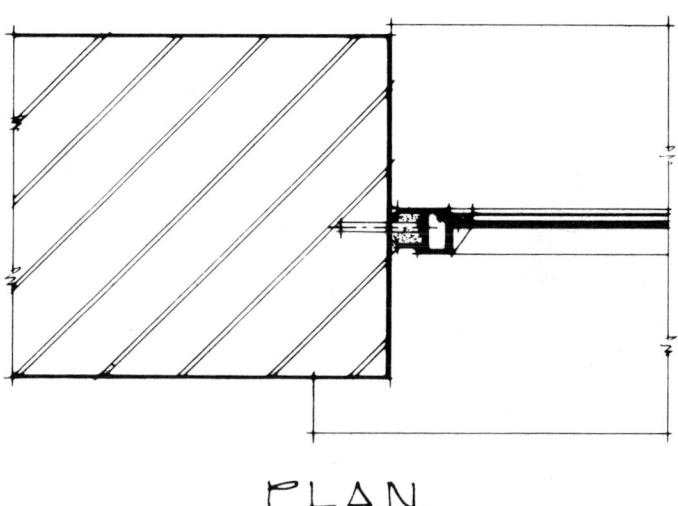

PLAN.   SCALE 1 : 5.

STANDARD GLASS SIZES   821 × 143mm. 848 × 666 mm.
PRIME AND TWO COATS OIL PAINT METAL WINDOW.
PRECAST CONCRETE 1:2:4 LINTEL AND SILL.

Plate 5
STOCK PATTERN WINDOW

S.M.M. RULES

Section A
F.18.21.
G.5.9.14.26.27.37.43.
N.1.13.14.17.21.31.
T.4.5.
U.2.3.4.
V.1.2.3.4.5.

APPROACH:-

Window   - Stock pattern window
                             Glass
                             Fixing
                             Painting

Opening
adjustment   - Walls & finishings
                             Head inside & out
                             Reveals inside & out
                             Cill inside & out.

(wdw

1	Stock Pattern casement wdw (magnet Joinery Type 240V) wi pressed steel butts, fastener & stays.		S.M.M. N17.21. A6. Stock pattern windows are supplied as complete units with opening casements already hung on butt hinges. Any ironmongery supplied with the window must be included in the description. All other ironmongery must be measured separately.

0.485×0.205
$= 0.099 \, m^2$

S.M.M. U2.3.4.

- Classification n.e. 0.10m²

0.49
0.21

3 mm thk clear
sheet glass &
glazg to wood
wi linseed oil
putty & sprigs
in panes n.e.
0.10 m² (In No.1 pane)

Description -
Kind &
nominal
thickness   - 3mm thick clear sheet
Compound   - linseed oil putty
Method   - putty
Securing   - sprigs
Sash   - wood
Classification-panes n.e. 0.10m²
No. of panes - one

| | | 0.545×0.815 = 0.444 m² | – | Classification 0.10 – 0.50m² |

| 0·55 0·82 | Ditto in panes 0.10 – 0.50 m² | | |

| | 0.485×1.045 = 0.507 m² | – | Classification 0.50 – 1.00m² |

| 0·49 1·05 | Ditto in panes 0·50 – 1.00 m² | | |

| 2/ 2 | 25 × 3 mm Galv m. s. fixg cramps 150 mm gth one end turned up holed & scrd to fw. other end b.i. bkk. | | S.M.M. N31. |

| 4/ 1·20 | Red wood frame & ptd one side in mastic | | S.M.M. G43. |

| | | | S.M.M. V5. The glass size classifies size of pane. |

| | 0.099 m² 0·444 m² 0·507 m² 3) 1·050 = 0·350 m² = MEDIUM PANES | – | Pane size averaged as SMM V5. |

1·23	
1·23	

K.P.S. ② wood
wdw in medium
panes

(new wk
  INT.

&

Ditto   (new wk
         EXT.

This item includes painting on ironmongery (S.M.M. V5).

S.M.M. V1.2.3.5.

```
         glass   0·485
Sash ⅖·035=  0·070
Casemt W =   0·555
             0·205
             0·070
Casemt H =   0·275
             0·555
             0·275
          2/ 0·830
             1·660
```

```
         glass   0·485
             0·070
Casemt W =   0·555

         glass   1·045
Sash top     0·035
Sash bot     0·050
Casemt H =   1·130
             0·555
             1·130
          2/ 1·685
             3·370
```

1·66	
3·37	

Ditto on
edges opg
casement

(new wk
  EXT.

— Opening edge includes additional painting on the surrounding frame.

40

(Adjust:

1·23 1·23	**Ddt** H.B. skin hollow wall entirely facgs ab &		S.M.M. G5.14. Deduct overall size of window.

Ddt Facing cavity ab &

S.M.M. G9.

Ddt 100 mm thk conc block skin hollow wall ab &

S.M.M. G27.

Ddt Two ct plaster ab. walls &

S.M.M. T4.5.

Ddt 2ce emulsion ditto

S.M.M. V1.2.3.

```
               1·225
 BI ³/₆·₁₀₀ 0·200
               1·425
```

S.M.M. F18.21.
Particulars of materials given in Preamble in preference to description.

Precast conc (1:2:4) bed lintol size 1425 × 260 × 150 mm fin fair 110 mm gth reinf wi 2 No 12 mm Ø m. s. bars wi 2 No notched ends to nib wi blockwk in mor (1:1:6).

S.M.M. F18.   Description –
Mix   – precast conc. (1:2:4)
Reinf.   – Two 12mm dia. m.s. bars.
Surf. finish – fin. fair 110mm gth.
Bedding   – mortar (1:1:6)
Fixing   – built into blockwork

Description quite adequate – bill diagram not required.

S.M.M. F3.
Classification – other concrete work.

No deduction made on inner skin – full block course not displaced – Block course 225mm high. Lintol 150mm high – SMM G26.

41

1·23	_Ddt·_		No deduction made for ends of lintol.
0·08	Forming cavity ab		S.M.M. G48.
	&		

	Ddt·	
	H.B. skin hollow	
	wall entirely	
	fегs ab	S.M.M. G37.

	Lintel	1·425
O'hang at		
Ends 2/0·100		0·200
	L =	1·625
	H/B	0·130
	CAVITY	0·090
	B.I	0·030
	W =	0·250

1·63	One layer hessian		Description –
0·25	based bituminous		Kind &
	felt D.P.C. forming		quality – hessian based bit. fel
	cavity gutter in		Substance – 3.8kg/m²
	hollow walls		Layers – one
	wkg 3·8 kg/m²		Bedding – mortar (1:1:6)
	bedd in mor (1:1:6)		Laps – no allowance
	no allowance		Classification – cavity gutter in
	made for laps		hollow wall

2	Ends		

			S.M.M. T4.
3/ 1·23	Two ct plaster		
0·15	a.b. walls n.e.		
	300 mm wide		
	&		
	Tce emulsion		S.M.M. V3.4.
	ditto.		Not isolated surface.
			S.M.M. T5.
3/1·23	Extl angle plaster		

2/1.23	Closing cavity 50mm wide at jambs of opgs wi blockwk incl one layer hess. based bit. felt vertical D.P.C. whg 3.8 kg/m² 100 mm wide bedd in mor (1:1:6) no allowce made for laps.	S.M.M. G9.	

$$\&$$

Fair return fcg bks H.B. wide

S.M.M. G14.
External reveal.

1.23	200 × 20 mm S.W. rebated wdw bd. wi rndd edge plugged to blockk	S.M.M. N13.14. Joinery labours appendix B.

2	Ends	S.M.M. N1. Over 0.002m² cross-sectional area

	0.190	S.M.M. V3.5. Not isolated surface.
	0.020	
	0.025	
	0.235	

1.23	K.P.S. ②	
0.24	wood wdw in medium panes (new wk (INTL.	

43

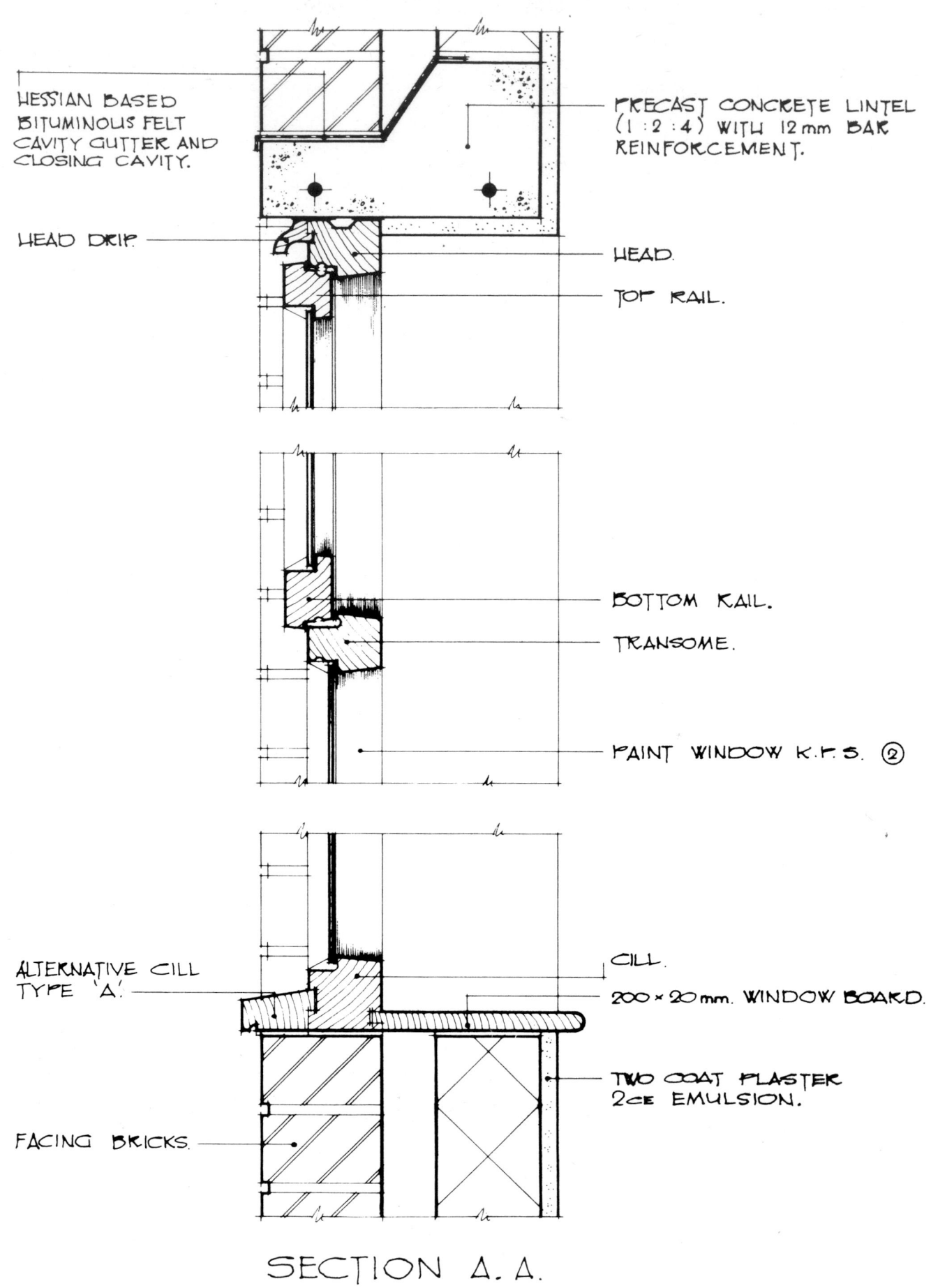

HESSIAN BASED
BITUMINOUS FELT
CAVITY GUTTER AND
CLOSING CAVITY.

PRECAST CONCRETE LINTEL
(1 : 2 : 4) WITH 12 mm BAR
REINFORCEMENT.

HEAD DRIP.

HEAD.

TOP RAIL.

BOTTOM RAIL.

TRANSOME.

PAINT WINDOW K.P.S. ②

ALTERNATIVE CILL
TYPE 'A'.

CILL.

200 × 20 mm. WINDOW BOARD.

TWO COAT PLASTER
2ce EMULSION.

FACING BRICKS.

SECTION A. A.

STANDARD GLASS SIZES :
485 × 205 mm.
545 × 810 mm.
485 × 1045 mm.

SCALE.   1 : 5.

Plate 5
STOCK PATTERN CASEMENT WINDOW
MAGNET JOINERY TYPE 240V

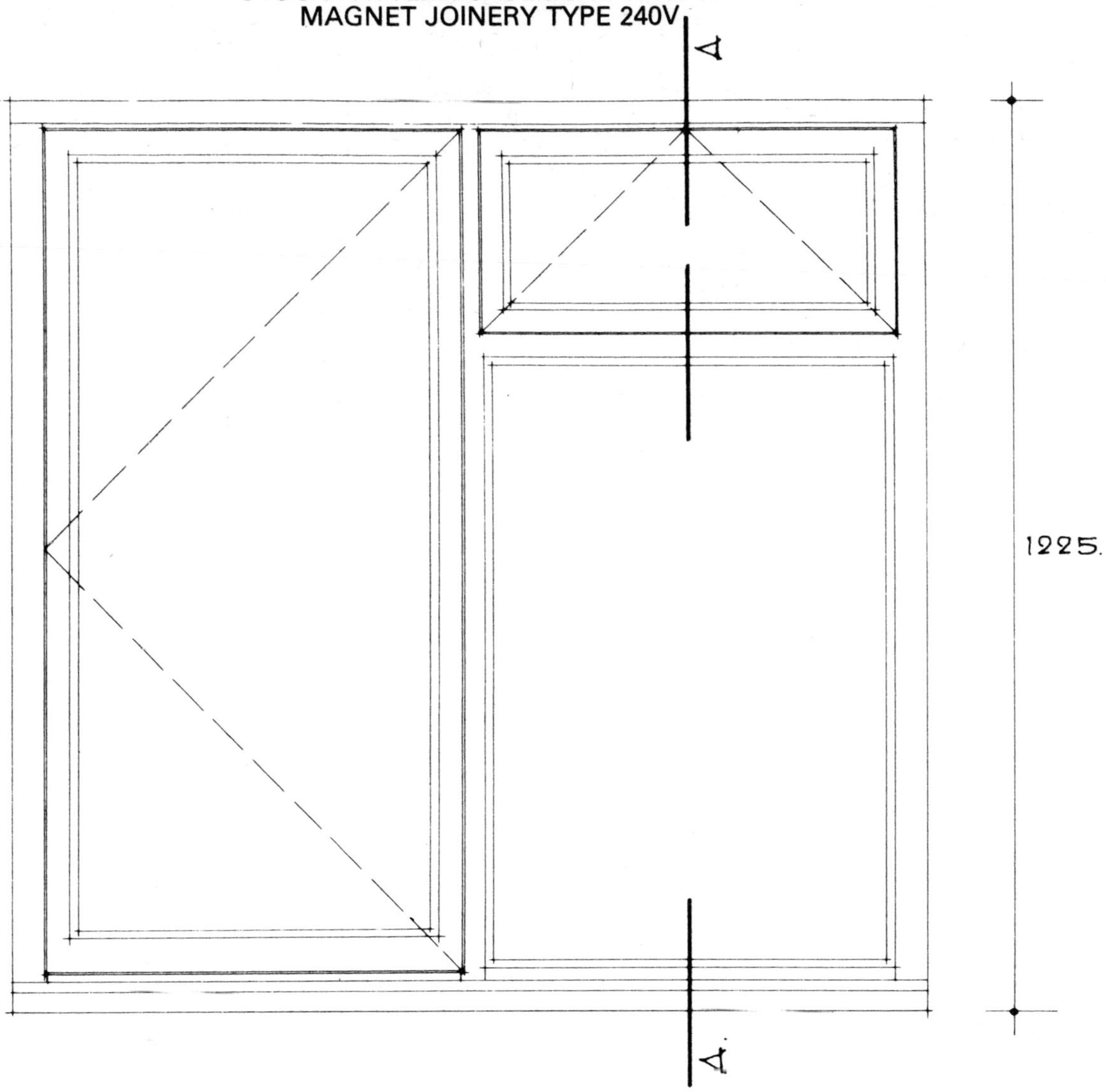

A

1225.

A.

ELEVATION.

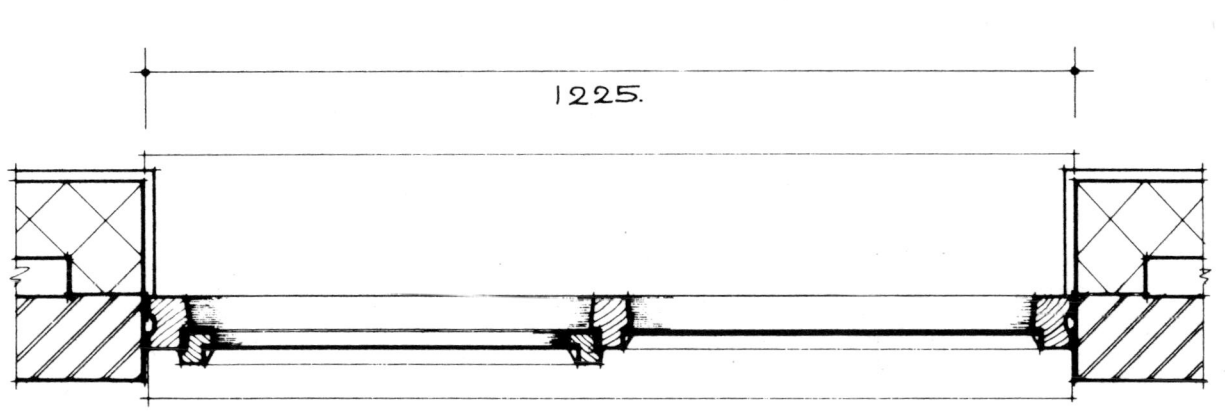

1225.

PLAN.

SCALE 1 : 20.

Plate 6
CASEMENT WINDOW

S.M.M. RULES.   SECTION A
                        F.18.21.
                        G.5.9.14.20.37.43.
                        N.1.13.14.17.21.31.32.
                        T.4.5.
                        U.2.3.4.
                        V.1.2.3.4.5.12.

APPROACH:-   Window   - Window
                                Ironmongery
                                Glass
                                Fixing
                                Painting
                    Opening
                    adjustment - Walls & finishings
                                Head inside & out
                                Reveals inside & out
                                Cill inside & out.

BILL DIAGRAM No 10   S.M.M. N1.17.21.

Joinery labours appendix B.

100×50mm FRAME, REBATED, CHECK THROATED, GROOVED & SPLAYED.
100×50mm MULLIONS, 2ce REBATED, 2ce CHECK THROATED, 2ce SPLAYED.
100×50mm TRANSOMES, 2ce REBATED, 2ce CHECK THROATED, 2ce SPLAYED.
200×75mm CILL, SUNK WEATHERED, CHECK THROATED, THROATED,
        SPLAYED & GROOVED
50mm THICK CASEMENTS, REBATED & THROATED
1 PAIR 75mm STEEL BUTTS TO EACH OPENING CASEMENT.

	1	West s.w. casement window size 2700 x 1500 mm o/a as Bill Diagram No 10		<u>S.M.M. N1.17.21.</u> Particulars of quality given in Preamble in preference to description.

<u>S.M.M. N32.</u>

2/ 2   250 mm long silver Anodised aluminium cast. stay Paker Winder Achurch Ltd B'ham Ref OP 10869 screwd to s.w.

–   Appropriate reference from trade catalogue.

2   Silver anodised aluminium cast. fastener P.W.A. Ltd Ref OP 10870 ditto

–   <u>S.M.M. N32.</u> Side hung casements only.

<u>S.M.M. U2.3.4.</u>

–   Top casement

Casement 0.600
Sash 2/0.040 <u>0.080</u>
Glass W = <u>0.520</u>

Casmt 0.370
       0.080
Glass H = <u>0.290</u>

$0.520 \times 0.290 = \underline{0.151 \, m^2}$

–   Classification 0.10 – 0.50m²

–   Bottom casement – width as for top casement.

Glass W = <u>0.520</u>
Wdw Ht = 1.500
head 0.040
Casmt 0.370
transom 0.030
cill <u>0.045 = 0.485</u>
Casmt = 1.015
Sash 2/0.040 = <u>0.080</u>
glass H = <u>0.935</u>

$0.520 \times 0.935 = \underline{0.486 \, m^2}$

–   Classification 0.10 – 0.50m²

Wdw L = 2.700   —   Top centre pane.

Casmts
   2/0.600 = 1.200
frame
   2/0.040 = 0.080
mullions
   2/0.030 = 0.060 = 1.340
     glass W = 1.360
     glass H = 0.370   —   Height as top casement.
1.360 × 0.370 = 0.503 m²   —   Classification 0.50 – 1.00m²

     glass W = 1.360   —   Bottom centre pane – width as top centre pane.
     glass H = 1.015   —   Height as bottom casement.
1.360 × 1.015 = 1.380 m²   —   Classification over 1.00m²

Description –

Kind & thickness	–	4mm Thk. clear sheet
Compound	–	linseed oil putty
Method of glazing	–	putty
Method of securing	–	sprigs
Sash	–	wood
Classification	–	various.

2/0.52
0.29

2/0.52
0.94

4 mm Thk clear sheet glass & glazing to wood wi linseed oil putty & sprigs in panes 0.10 – 0.50 m²

1.36
0.37

Ditto in panes 0.50 – 1.00 m²

1.36
1.02

Ditto in panes over 1.00 m²

2/3

25 × 3 mm galv M.S. fixg cramp 150 mm gth one end turned up holed & scrd to stw other end b.i. bkk.

     S.M.M. N31.

2/ 2·70 2/ 1·50	Bed wood frame & pt one side in mastic	S.M.M. N31.	

$$2/0.151 = 0.302$$
$$2/0.486 = 0.972$$
$$0.503$$
$$1.380$$
$$6)\overline{3.157}$$

Avg. $\underline{0.526\,m^2}$ – average large panes.

S.M.M. V1.2.3.5.
Glass size classifies size of pane.

2·70 1·50	K.P.S. ③ wood wdw in Avg large panes & (new wk Intl. ditto (new wk Extl.	Particulars of materials given in Preamble in preference to description.	

2/ 2/ 2/ 0.60 2/2/ 0.34 2/2/ 1.02	ditto edges of opg cast (new wk Extl. (Opening Adjust	S.M.M. V5. Outer edges of casements.	

		Description		S.M.M. Notes

Ddt:
HB skin hollow wall entirely fcgs ab
&

2.40
1.50

S.M.M. G14.
Deduct overall size of window.

Ddt:
Forming cavity ab
&

S.M.M. G9.

Ddt:
HB skin hollow wall in cb ab
&

S.M.M. G5.

Ddt:
Two ct plaster a.b. walls
&

S.M.M. T5.

Ddt:
Hang pattern wall paper ab.

S.M.M. V12.

BI   2.400
2/0.100   0.200
              2.900

1

Precast Conc (1:2:4) lintol 2900 x 150 x 150 mm all keyed for plaster reinf wi 2/ 12mm ø m.s. bars & b.i. bkt in mor (1:1:6).

S.M.M. F18.21.
Particulars of material given in Preamble in preference to description.

Description quite adequate – bill diagram not required.

S.M.M. F3.
Classification – other concrete work.

2.40
0.15

Ddt
Forming cavity ab
&
Ddt:
HB skin hollow wall in cb ab.

No deduction made for ends of lintel S.M.M. G48.

50

Lintel 2·900

O'hang at
ends 2/0·100   0·200

$\qquad$ L=   3·100

B.I.   0·030

Cavity   0·250

head   0·050

W =   0·330

```
3·10
0·33
────
```

One layer hessian
based bit. felt
D.P.C. forming cavity
gutter in hollow
walls wkg 3·8 kg/m²
bedd in mor (1:1:6)
no allowce made
for laps

```
2
────
```

Ends

```
2·40
────
```

Brick on end flat
arch entirely fegs
ab. 225 mm wide
on face 75 mm wide
on exposed soffit
in mor (1:1:6) ptd
wi neat w/s jt

S.M.M. G20.
Particulars as S.M.M.G20 but
brickwork entirely facing bricks
NOT E.O. facework.

```
2·40
0·23
────
```

Ddt
H.B. skin hollow
wall entirely
fegs ab

S.M.M. G14.

	2·40	Two ct. plaster	S.M.M.	T4.5.	
	0·10	a.b. walls n.e.			
2/	1·50	300 mm wide			
	0·10	&			
		Hang pattern	S.M.M.	V12.	
		wallpaper a.b			
	2·40	Extl angle	S.M.M.	T5.	
2/	1·50	plaster			
2/	1·50	Closing cavities	S.M.M.	G9.37.	
		50 mm wide at			
		jambs of opgs			
		wi bkk incl one			
		layer hessian			
		based bit. felt			
		vertical D.P.C. whg			
		$3.8 \, kg/m^2$ 112·5 mm			
		wide bedd in			
		mor (1:1:6) no			
		allowce made for			
		laps			
		&			
		fair return fcg	S.M.M.	G14.	
		bks HB wide		External reveals	
	2·40	140×20 mm Wrot	S.M.M.	N13.	
		sw rebated wdw			
		bed wi rndd edge			
		pl. to bkk			
		Ends	S.M.M.	N1.	
2/	1				
	2·40	K.P.S. ③ wood wdw	S.M.M.	V3.4.5.	
	0·16	ang large panes			
		(new ink			
		note.			

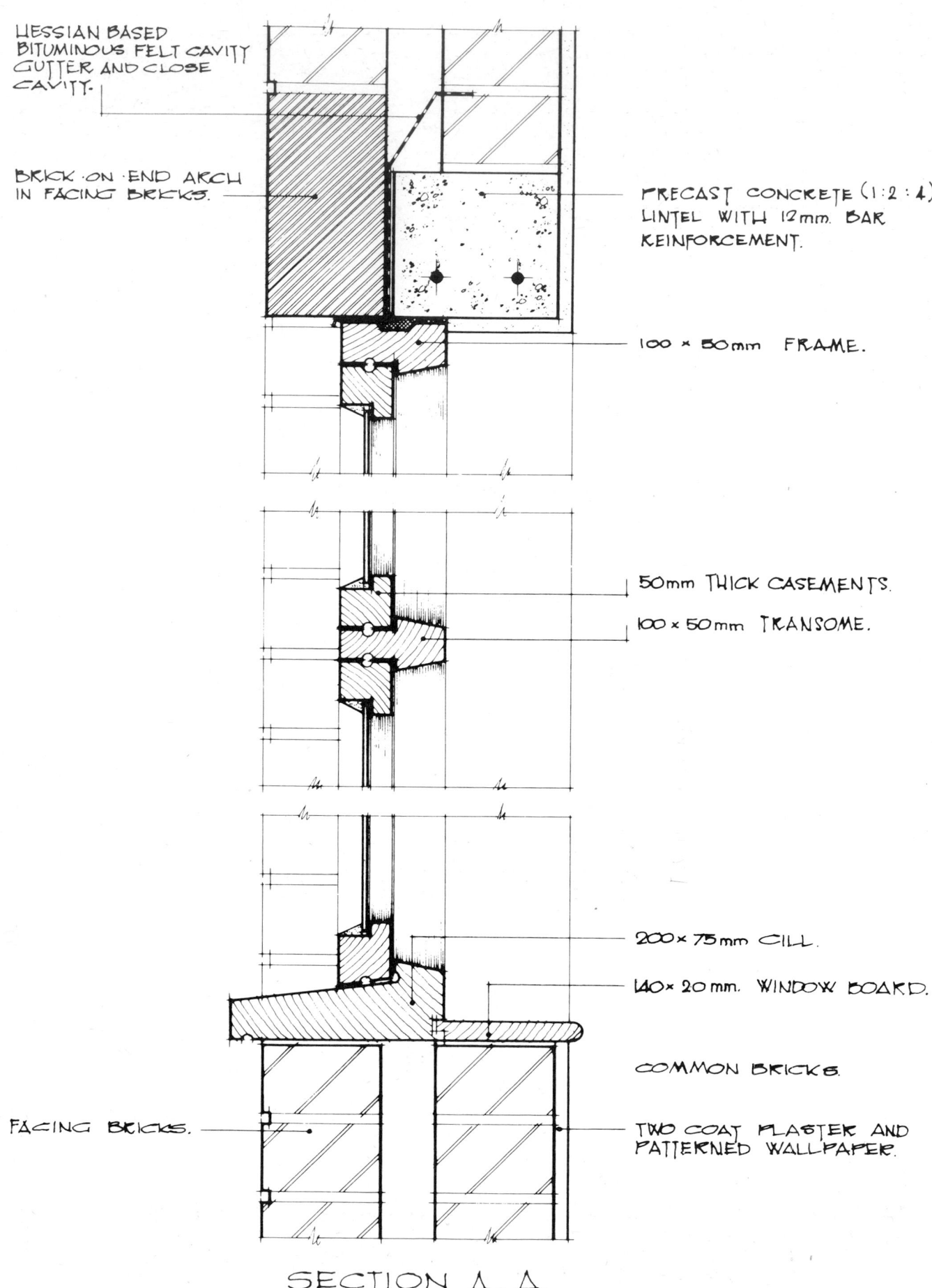

HESSIAN BASED
BITUMINOUS FELT CAVITY
GUTTER AND CLOSE
CAVITY.

BRICK ON END ARCH
IN FACING BRICKS.

PRECAST CONCRETE (1:2:4)
LINTEL WITH 12mm. BAR
REINFORCEMENT.

100 × 50mm FRAME.

50mm THICK CASEMENTS.

100 × 50mm TRANSOME.

200 × 75mm CILL.

140 × 20mm. WINDOW BOARD.

COMMON BRICKS.

FACING BRICKS.

TWO COAT PLASTER AND
PATTERNED WALLPAPER.

SECTION A. A.

Plate 6
CASEMENT WINDOW

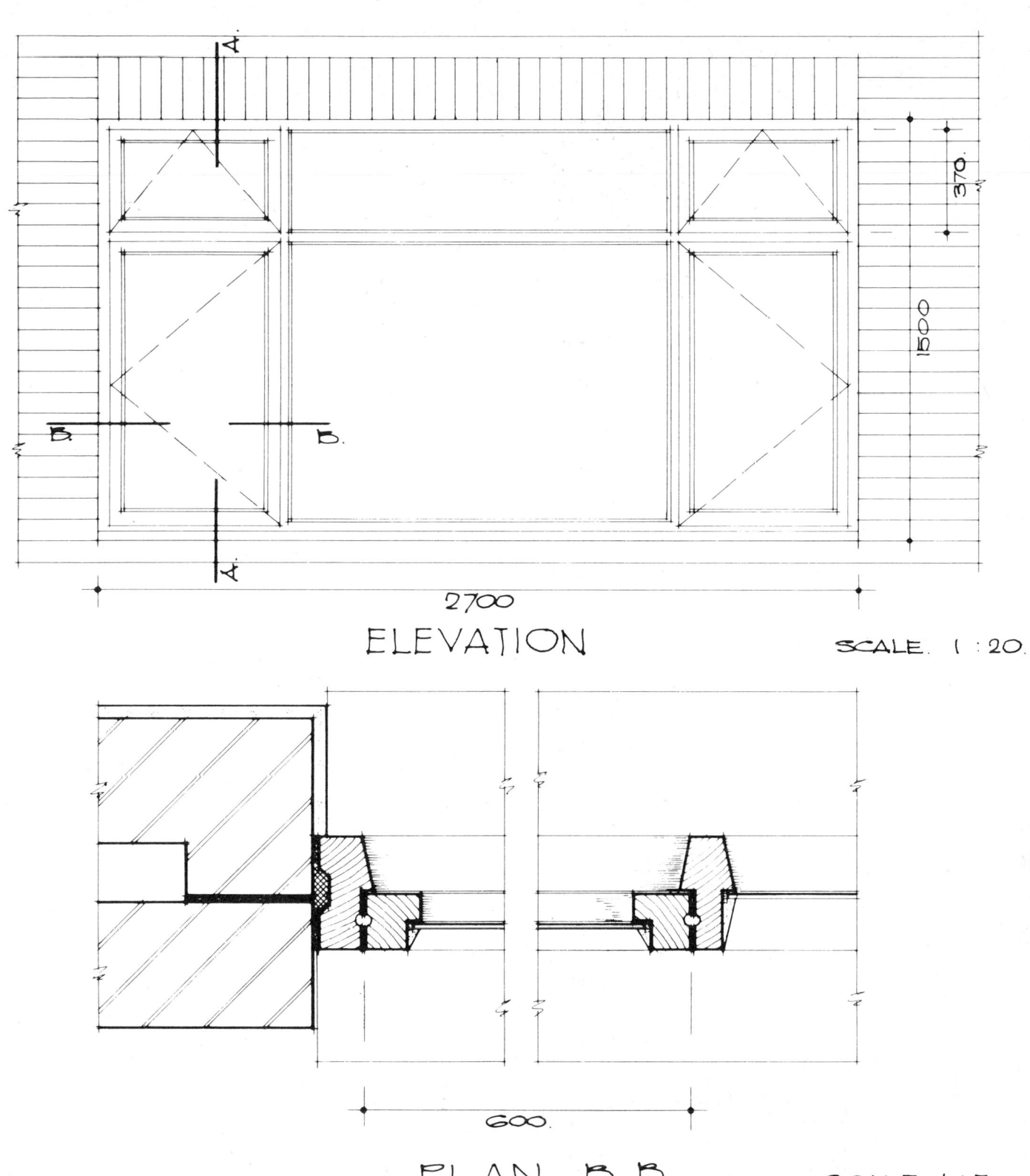

ELEVATION   SCALE. 1 : 20.

PLAN. B.B.   SCALE. 1 : 5.

IRONMONGERY :   75mm. STEEL BUTTS
PARKER · WINDER · ACHURCH LTP 250mm SILVER ANODISED
ALUMINIUM CASEMENT STAY Ref. OP 10869.
CASEMENT FASTENER   Ref. OP 10870.
4mm. THICK CLEAR SHEET GLASS.
PAINT WINDOWS K.F.S. ③.

Plate 7
PIVOTED CASEMENT

S.M.M. RULES

SECTION A
F.18.21.
G.5.9.14.27.43.
N.1.17.21.
Q.2.
T.13.14.

APPROACH:—

Window — Window
Ironmongery
Glass
Fixing
Painting

Opening
adjustment — Walls & finishings
Head inside & out
Reveals inside & out
Cill inside & out.

_____

(window

BILL DIAGRAM No 11.   S.M.M. N21.

900   900

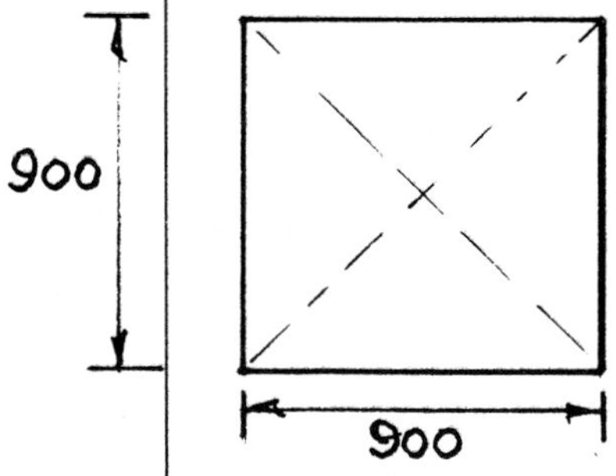

Joinery labours appendix B.

50 mm THICK CASEMENT, REBATED, THROATED & SPLAYED BOTTOM EDGE.
100 x 50 mm HEAD, SPLAYED & THROATED.
100 x 50 mm JAMBS, THROATED.
200 x 75 mm CILL, SUNK WEATHERED, CHECK THROATED, THROATED.
25 x 19 mm CUT & MITRED STOP FILLETS AROUND CASEMENT.
PAIR B.B. FRICTION PIVOTS. P.W.A. LTD REF OP 10951
C.P. FANLIGHT CATCH. P.W.A. LTD. REF S.P. 11102.

1

Softwood
Pivoted casement
wdw size 900 x 900 mm
as Bill Diagram
No 11.

S.M.M. N21.17.
Particulars of quality given in
Preamble in preference to description.

			Casement 0·800	S.M.M. U2.3.4.	

sash

²/0·040   0·080

glass W = 0·720

Casement 0·795

sash 0·035
   0·040 0·075

glass H = 0·720

0·720 × 0·720 = 0·518 m² —   Classification 0.50 – 1.00m²

0·72	
0·72	

4 mm Clear sheet glass & glazing to wood wi linseed oil putty & sprigs in panes 0.50 – 1.00 m²

²/ 2   S.M.M. N31.

25 × 3 mm Galv
m.s. feg cramp
150 mm gth one
end turned up,
holed & scrd to
ther other end
bi. blockwk

4/0.90   S.M.M. G43.

Bed wood
frame & ft
one side in
mastic

0.90 0.90	K.P.S. ③ wood wdw in large panes ⟮new wk Intl. $\cancel{}$ Ditto ⟮new wk Extl.	S.M.M. V1.2.3.5. Particulars of materials given in Preamble in preference to description. Dimensions as window. Glass size classifies size of pane.	

frame²⁄0.050 $\dfrac{\begin{array}{r}0.900\\0.100\\\hline 0.800\end{array}}{}$

S.M.M. V5.

4/0.80	Ditto edges of opg casement ⟮new wk Extl. ⟮Opening ⟮Adjust-	S.M.M. G5.14. Deduct overall size of window.
0.90 0.90	Ddt. HB skin hollow wall entirely fcgs ab $\cancel{}$	
	Ddt. Formg cavity ab $\cancel{}$	S.M.M. G9.
	Ddt. 100 mm Thk block skin hollow wall ab. $\cancel{}$	S.M.M. G27.
	Ddt. 15 mm Thk floated backg ab $\cancel{}$	S.M.M. T13.
	Ddt. Cream glazed wall tiles ab	S.M.M. T14.

		$BI \, ^2/0.100 \, \begin{matrix} 0.900 \\ 0.200 \\ \hline 1.100 \end{matrix}$	S.M.M. F18.21.

S.M.M. F18.21.

Particulars of material given in Preamble in preference to description.

Description quite adequate – bill diagram not required.

1

Precast
Conc (1:2:4) lintel
size 1100 x 100 x 150 mm
all keyed for
plaster reinf wi
1 No 12 mm φ M.S.
bar b.i. blockk
in mor (1:1:6)

S.M.M. F3.
Classification – other concrete work.

No deduction made on inner skin.
Full block course not displaced.
Block course 225mm high, lintel
150mm high. S.M.M. G26.

$BI \atop ^2/0.150 \, \begin{matrix} 0.900 \\ 0.300 \\ \hline 1.200 \end{matrix}$

S.M.M. A6. Q2.
Standard product.

1

Galvd. steel
standard lintel
1200 mm long
152 mm deep
whg 7.4 kg/m
Dorman Long (Steel)
Ltd., middlesbrough
& b.i. blockk in
mor (1:1:6)

$4/0.90 \atop 0.10$

ct & sand (1:3)
15 mm thk
floated backg
n.e. 300 mm
wide on blockk.

S.M.M. T13.
Measured the area in contact with
the base. S.M.M. T4.

	$4/0.900 = 3.600$	
	backg 0.015	
	tiles $\underline{0.006}$	
	$4/2/$ $\underline{0.021}$ 0.168	
	$\underline{3.432}$	

3.43	cream glazed	
0.11	wall tiles a.b.	
	n.e. 300 mm wide	

3.43	E.O. rounded	S.M.M. T14.
	edge tile (R.E.).	

2/0.90	Close cavities	S.M.M. G9.
	50 mm wide at	
	jambs of opgs	
	wi blockk incl	
	one layer hessian	
	based bit. felt	
	vertical D.P.C.	
	whg 3.8 kg/m	
	100 mm wide	
	bedd in mort	
	(1:1:6) no allowce	
	made for laps.	

3/0.90	Fair return fcg	S.M.M. G14. External soffit and reveals.
	bks H.B. wide.	

60

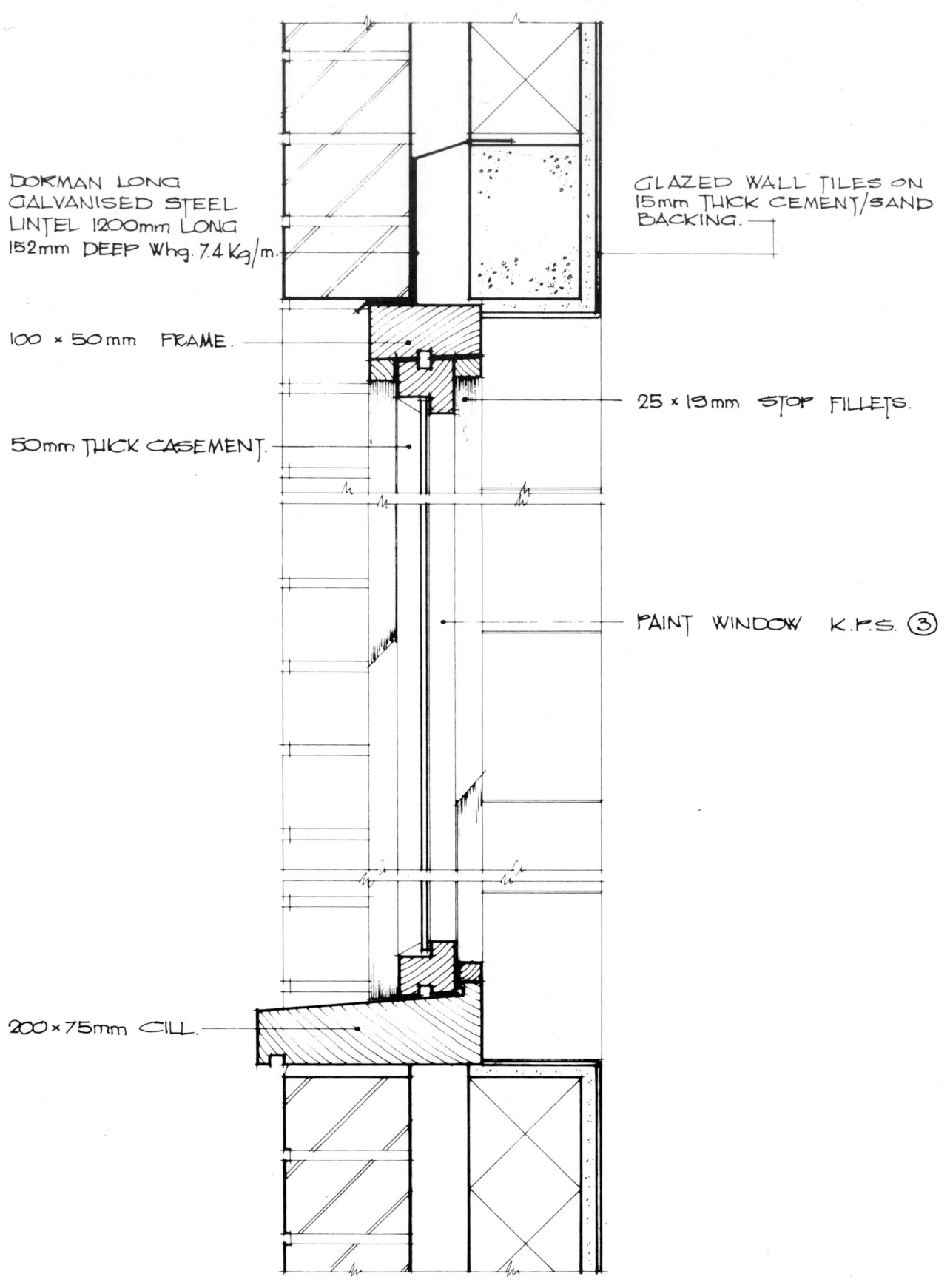

DORMAN LONG
GALVANISED STEEL
LINTEL 1200mm LONG
152mm DEEP Whg. 7.4 Kg/m.

GLAZED WALL TILES ON
15mm THICK CEMENT/SAND
BACKING.

100 × 50mm FRAME.

25 × 19mm STOP FILLETS.

50mm THICK CASEMENT.

PAINT WINDOW K.P.S. ③

200×75mm CILL.

SECTION A.A.

IRONMONGERY :
PAIR BERLIN BLACK FRICTION PIVOTS (P.W. A. LTD.) REF. OP 10851.
Nº I CHROMIUM PLATED FANLIGHT CATCH (P.W. A. LTD. REF. SP 11102.

SCALE 1 : 5.

Plate 7
PIVOTED CASEMENT

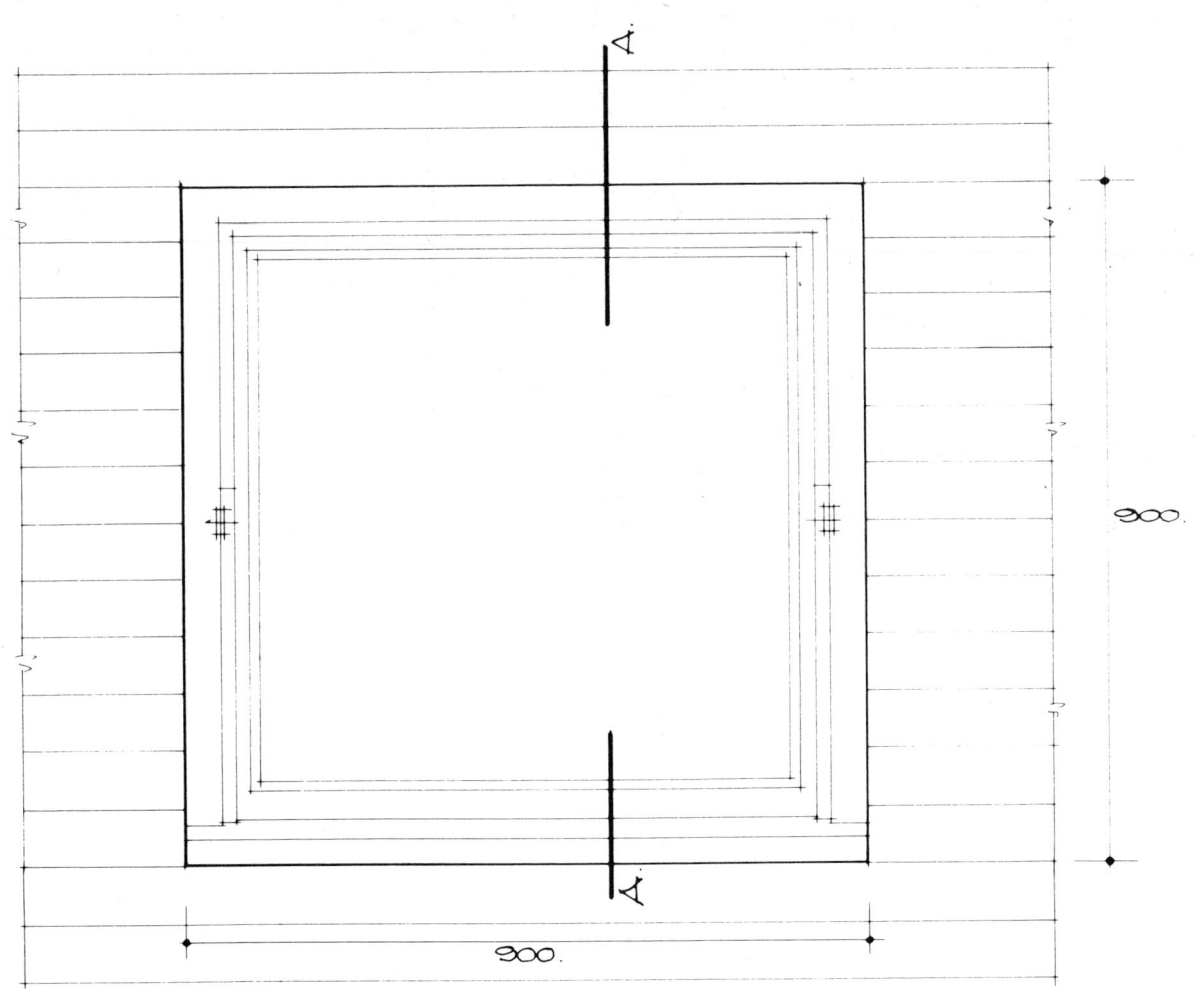

A

900.

A

900.

ELEVATION

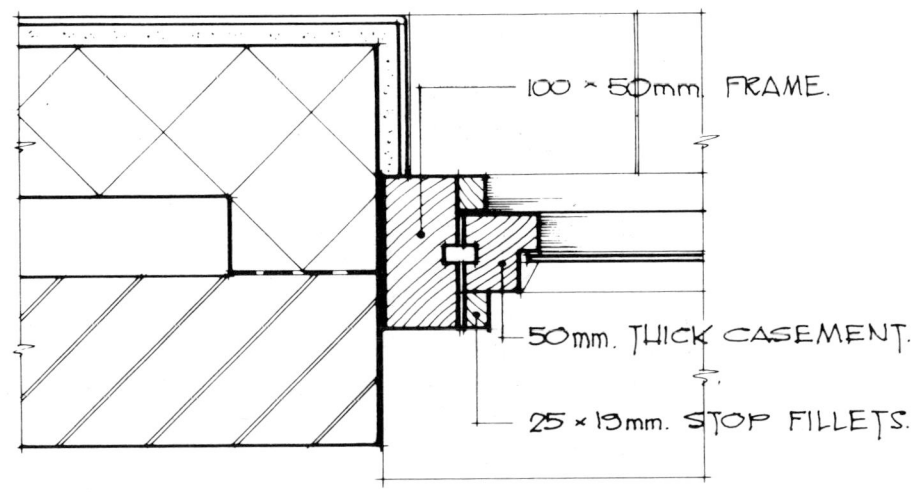

100 × 50mm. FRAME.

50mm. THICK CASEMENT.

25 × 19mm. STOP FILLETS.

PLAN.

SCALE. 1 : 10.

Plate 8
SASH WINDOW

S.M.M. RULES   SECTION A.
                              F.18.21.
                              G.5.9.27.37.43.48.
                              N.1.13.14.21.31.32.
                              Q.7.
                              T.4.5.
                              U.2.3.4.
                              V.1.2.3.4.5.

APPROACH:-   Window   - window
                                         ironmongery
                                         glass
                                         fixing
                                         painting.
                              Opening
                              adjustment -Walls & finishings
                                         Head inside & out
                                         Reveals inside & out
                                         Cill inside & out.

                              ───────────────

                                    (Window

BILL DIAGRAM   No 12.

<u>S.M.M. N1.21.</u>
Particulars of quality given in
Preamble in preference to description.

Joinery labours appendix B.

1500

1000

125×38 mm HEAD, GROOVED. & 125×38 mm JAMBS 2ce GROOVED
50×25 mm SPLAYED OUTSIDE LINING.
25×16 mm PARTING BEAD. 19×19mm SPLAYED INSIDE FILLET.
150×63 mm CILL, SPLAYED, SUNK WEATHERED, ROUNDED, THROATED &
THREE TIMES GROOVED.   50×19mm DRAUGHT BEAD, REBATED, SPLAYED.
44 mm THICK SLIDING SASHES WITH EXTRA DEEP BOTTOM RAIL, THROATED.
IRONMONGERY — P.W.A. LTD. TYPE M UNIQUE SPIRAL SPRING BALANCES
REF OP11042. BRASS SASH FASTENER REF SP10983. TWO BRASS SASH
PULLS REF SP11012. TWO BRASS FLUSH LIFTS REF SP10563.

1

West elv. double
hung sash window
size 1000×1500 mm
c/a as BILL
DIAGRAM No 12

W. 1.000
frames 2/0.038 0.076   S.M.M. U2.3.4.
            Sash 0.924
        2/0.040 0.080
    glass W = 0.844
            H = 1.500

frame 0.038
cill 0.060
top rail 0.040
meetg rail 0.030
bott rail 0.050 0.218
                2) 1.282

        glass H = 0.641

0.844 × 0.641 = 0.541 m²   —   Classification 0.50 – 1.00m²

2/ 0.84
   0.64

4 mm thick clear sheet glass & glazg to wood wi linseed oil putty & sprigs in panes 0.50–1.00m²

2/ 3

S.M.M. N31.

25 × 3 mm Galv m.s. fixg cramp 150 mm gth one end turned up holed & scrd to oth other end b.i. bkk

---

S.M.M. G43.

2/ 1.00
2/ 1.50

Bed wood frame & pt o/s in mastic

1·00	K.P.S. ③ wood	<u>S.M.M.</u> <u>V1.2.3.5.</u>	
1·50	wdw in large	Particulars of material given in	
0·92	panes (new wk	Preamble in preference to description.	
0·04	1 Nºr		
	&	The glass size classifies the size	
		of panes.	
	Ditto (new wk	Extra girth of edges S.M.M. V2.	
	13 × 72.	S.M.M. V5. for casements only.	
	(Opg		
	(adjust		
1·00	Ddt	<u>S.M.M.</u> <u>T5.</u>	
1·50	2 Ct Tyrolean	Deduct overall size of window.	
	rendering ab		
	&		
	Ddt		
	H. B. skin hollow	<u>S.M.M.</u> <u>G5.</u>	
	wall in c.b. ab		
	&		
	Ddt		
	Forming cavities	<u>S.M.M.</u> <u>G9.</u>	
	ab &		
	Ddt		
	100 mm Thk conc	<u>S.M.M.</u> <u>G27.</u>	
	block skin hollow		
	wall ab		
	&		
	Ddt		
	2 ct plaster ab	<u>S.M.M.</u> <u>T5.</u>	
	walls		
	&		
	Ddt		
	2cc emulsion	<u>S.M.M.</u> <u>V4.</u>	
	ditto		


```
                                    1·000
                        BI  2/0·100  0·200
                                    ─────
                                    1·200
             1      Precast conc (1:2:4)
                    boot lintol size
                    1200 x 300 x 150 mm
                    fin fair 150 mm
                    gth reinf wi
                    2 No 12 mm dia
                    m.s bars wi 2 No
                    notched ends to
                    nib & b.i.
                    blockk in mor
                    (1:1:6)


           1·00     Ddt
           0·08     Forwng cavities ab

                              &

                    Ddt
                    H.B. skin hollow
                    wall in cB ab

                              &

                    Ddt
                    Tyrolean rendering
                    ab


                          lintol   1·200
                       end
                       o'hang
                         2/0·100  0·200
                                 ──────
                             L = 1·400

                            nib  0·160
                         cavity  0·090
                           b.i.  0·040
                                ──────
                            W = 0·290
```

S.M.M. F18.21.

Particulars of material given in
Preamble in preference to description.

Description quite adequate —
bill diagram not required.

S.M.M. F3.
Classification — other concrete work.

No deduction made on inner skin.
Full block course not displaced.
Block course 225mm high, lintol
150mm high.   S.M.M. G26.

S.M.M. G9.
No deduction made for ends of
lintol. S.M.M. G48.

S.M.M. G5.

S.M.M. T5.

S.M.M. G37.

	1.40 0.29	One layer hessian based bit felt D.P.C. form'g cavity gutter in hollow walls wkg 3.8 kg/m² bedd in mor (1:1:6) no allowce made for laps		S.M.M. G37.
	2	Ends		
	1.00 0.07 2/1.50 0.07	Two coat plaster a.b. walls n.e. 300 mm wide &		S.M.M. T5.
		Dee emulsion ditto		S.M.M. V4.
	1.00 2/1.50	Extl angle plaster		S.M.M. T5.
	2/1.50	Closing cavities 50 mm wide at jambs of opg wi blockk incl one layer hessian based bit felt vert DPC wkg 3.8 kg/m² 112.5 mm wide bedd in mor (1:1:6) no allowce made for laps.		S.M.M. G9.37.
68				S.M.M. G37.

2/1.50 0.07	Two ct Tyrolean render a.b. walls n.e. 300 mm wide	<ins>S.M.M.</ins> T4.5.
1.00	100 × 25 mm shot dr reb wdw bd wi rnded edge plg to blockk	<ins>S.M.M.</ins> N13.
2	notched end size 20×70 mm	<ins>S.M.M.</ins> N14.
1.00 0.13	K.P.S. ③ wood wdw in large panes (newwk 1N12.	<ins>S.M.M.</ins> V3.4.5. not isolated surface.
1	BI 1.000 2/0.75 0.150 1.150 Precast conc (1:2:4) sunk weath, throated & grooved cill 1150×180×75 mm fin fair 200 mm gth wi 2 ½ stooled ends & b.i. lkk in mor (1:1:6)	<ins>S.M.M.</ins> F18.21.
1.00 0.08	<del>Ddt</del> Foring cavities ab & <del>Ddt</del> HB. skin hollow wall in cb ab & <del>Ddt</del> Two ct Tyrolean rendering ab	<ins>S.M.M.</ins> G9. <ins>S.M.M.</ins> G5. <ins>S.M.M.</ins> T5.
1.00	3 × 30 mm galv m.s. water bar in cill bedd in mastic.	<ins>S.M.M.</ins> Q7.

69

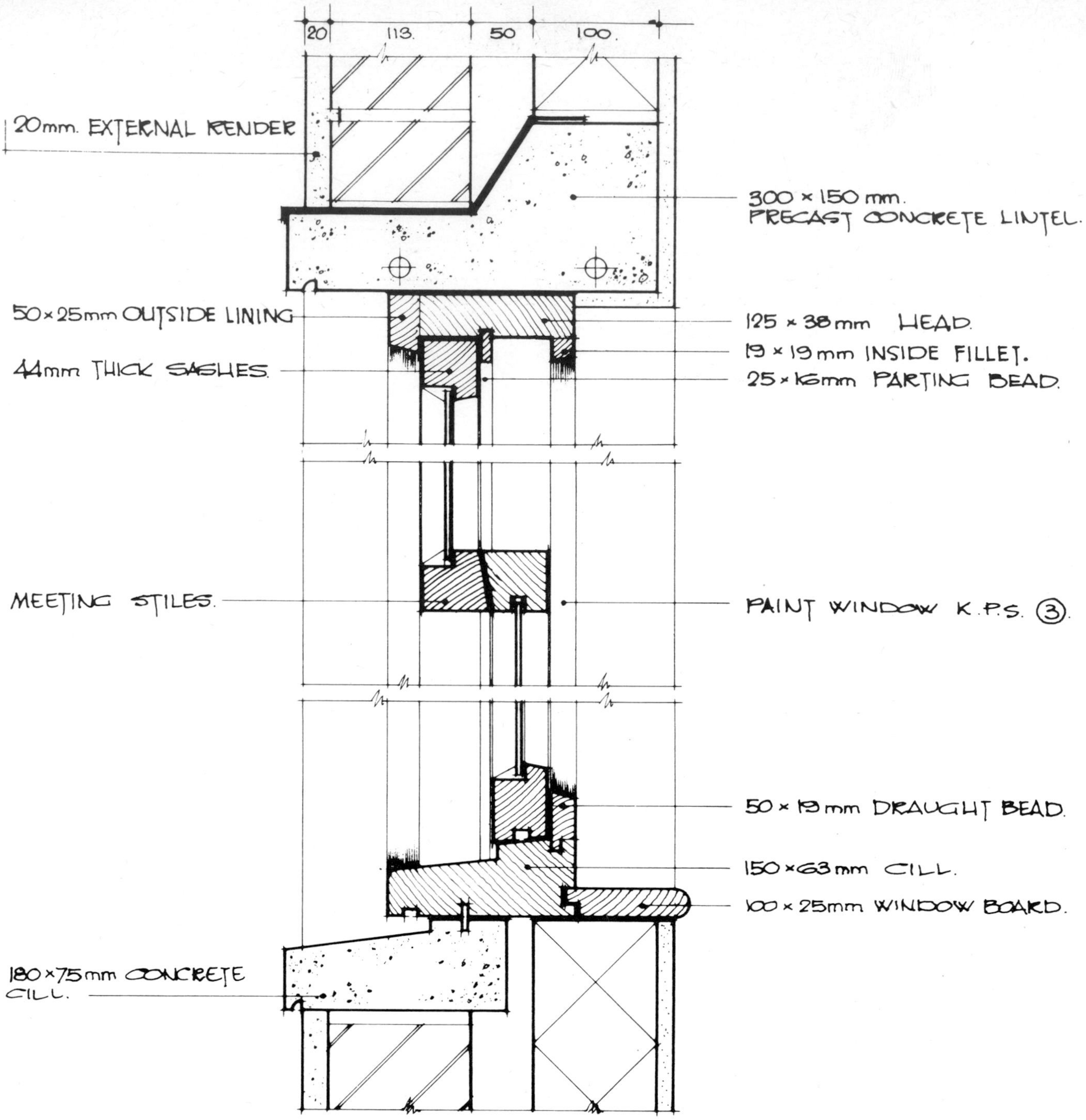

20   113.   50   100.

20mm. EXTERNAL RENDER

300 × 150 mm.
PRECAST CONCRETE LINTEL.

50 × 25mm OUTSIDE LINING

125 × 38mm   HEAD.

44mm THICK SASHES.

19 × 19 mm INSIDE FILLET.
25 × 16mm   PARTING BEAD.

MEETING STILES.

PAINT WINDOW K.P.S. ③

50 × 19mm DRAUGHT BEAD.

150 × 63mm CILL.

100 × 25mm WINDOW BOARD.

180 × 75mm CONCRETE
CILL.

SECTION A.A.

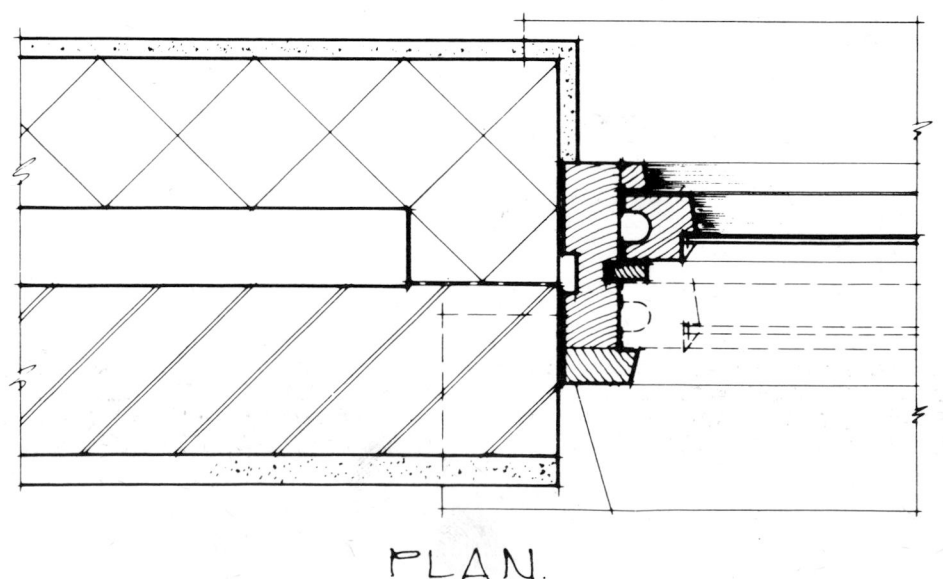

PLAN.

SCALE. 1 : 5.

Plate 8
SASH WINDOW

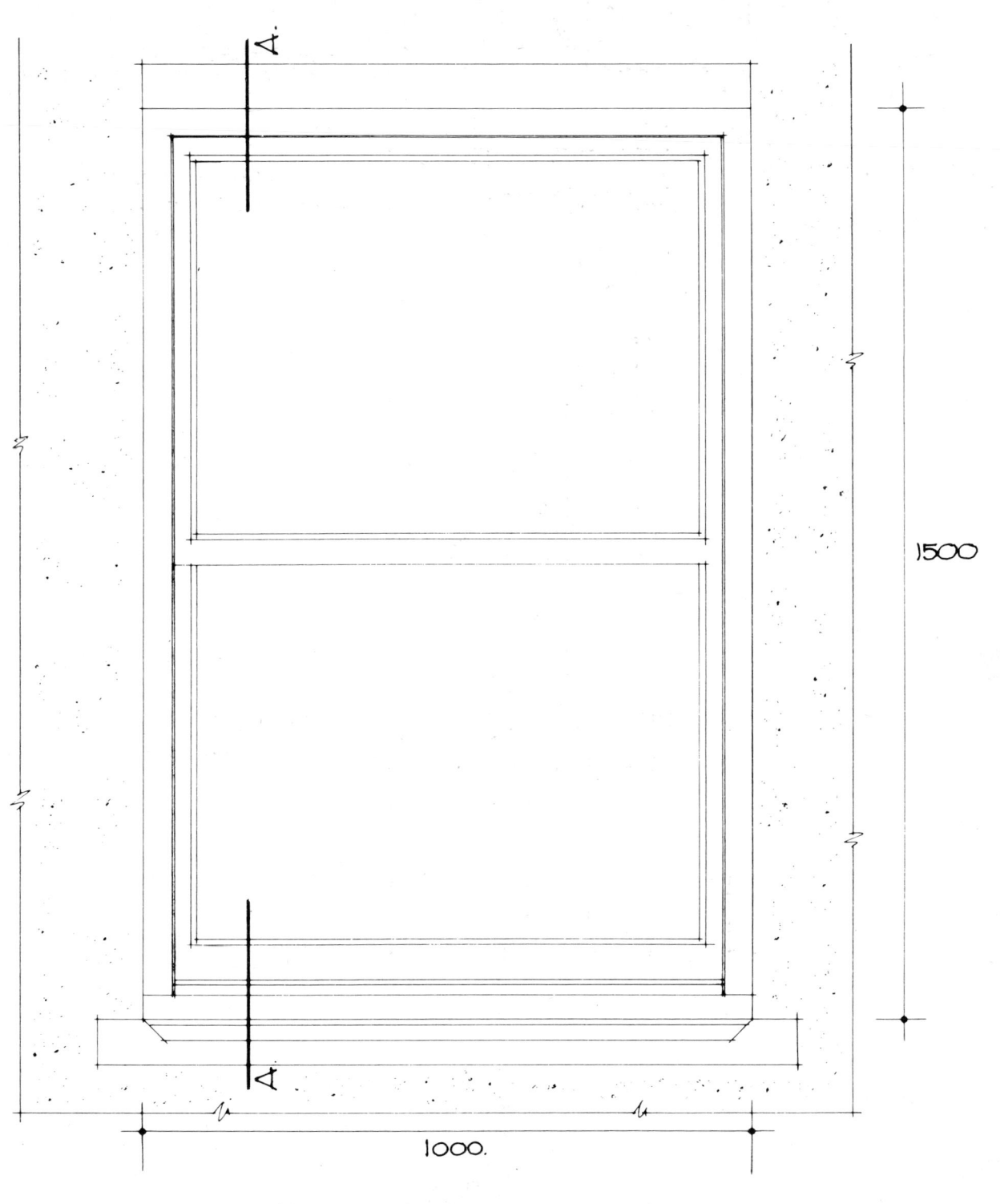

A.

1500

1000.

ELEVATION.   SCALE 1 : 10.

TYPE M UNIQUE SPIRAL SPRING BALANCE (F.W.A. LTD. Ref. OP 11042.
Nº 1 BRASS SASH FASTNER   (P.W.A. LTD. Ref. SF 10983.
Nº 2 BRASS SASH PULLS   (P.W.A. LTD. Ref. SF 11012.
Nº 2 BRASS FLUSH LIFTS   (P.W.A. LTD. Ref. SF 10563.

CHAPTER III

Doors

Approach

(1) Where a large project includes several buildings measure each as a separate unit.

(2) Compare the door schedule with the drawings to ensure that all doors have been given a reference number and are included in the schedule. Schedules provide the essential information in a concise, tabulated form for easy reference, as a more practical alternative to embodying the details throughout the drawings. This saves time involved in searching for information, and where schedules are not provided by the Architect the taker off is well advised to prepare his own. The preferred forms of schedule recommended by BS 1192 'Building Drawing Practice' are illustrated in Appendix C. Many surveyors prefer to prepare their own schedules giving particulars of doors and openings to enable them to measure direct from the schedules without further reference to the drawings.

The use of a schedule avoids unnecessary repetition when taking off, since doors with similar features may be readily identified and grouped together.

(3) The measurement of doors normally includes the adjustments of walls and finishings which have been measured over the door openings.

(4) Each door must be considered on its merits but a typical sequence of measurement would be:-

Door	door
	ironmongery
	glass
	frame
	fixing frame
	painting
Opening adjustment	walls and finishings
	head inside and out
	reveals inside and out
	threshold inside and out

Measurement of doors

METAL DOORS. Standard metal doors shall each be enumerated separately (SMM Q2) and (SMM A6) are normally identified by a reference number relating to the appropriate BS.

Special metal doors (complete with frames mullions, transomes, hinges and fastenings) shall each be enumerated separately stating the overall size, the nature of the construction, the shape where other than rectangular, the finish and the number of opening portions (SMM Q3).

Building in metal doors (complete with frames) shall be enumerated stating the overall size. Building in or cutting and pinning lugs, bedding frames and pointing to one or both sides shall be given in the description (SMM G48).

WOOD DOORS. Doors shall be described and enumerated (each leaf being counted as one door (SMM N19). Stock pattern doors may be identified by a reference number relating to the manufacturer's catalogue and standard doors by a reference to the BS. The taker off should ascertain which ironmongery is supplied with each door, since additional ironmonery must be measured separately in accordance with SMM N32.

Labours on doors (e.g. rebates and rounded heels) shall be given in the description of the door (SMM N17). Fitting and hanging shall be deemed to be included with the items (SMM N19). Glazing beads and glazing fillets may be given in metres (SMM N13) or included in the description of the appropriate door (SMM N17).

FRAMES. Steel door frames (except those forming an integral part of a door unit) shall be enumerated stating the size (SMM Q3). Building in or cutting and pinning lugs, bedding frames and pointing to one or both sides shall be given in the description (SMM G48).

Wood door frame and lining sets shall be grouped together stating the number of sets. Jambs, heads, cills, mullions, transomes and the like shall each be fully described and given separately in metres stating the cross-section dimensions and the number of lengths of cills, mullions and transoms. Attention shall be drawn to the incidence of repetition of identical items (SMM N20). In determining the incidence of repetition, it is important to consider the number of different cross-section shapes (SMM N1) e.g. jambs of identical size with rebates of differing size are not identical items. Wood door frames, cills, mullions and transoms are measured the length as fixed in the work (SMM A3). Thus when the horns are not cut off whilst fixing the frame, the actual length of the horns is included in the length as fixed in the work; when horns are cut off before fixing no allowance is made. The length of frames, mullions, transoms includes all tenons and therefore framed ends, angles, intersections and the like are deemed to be included (SMM N17).

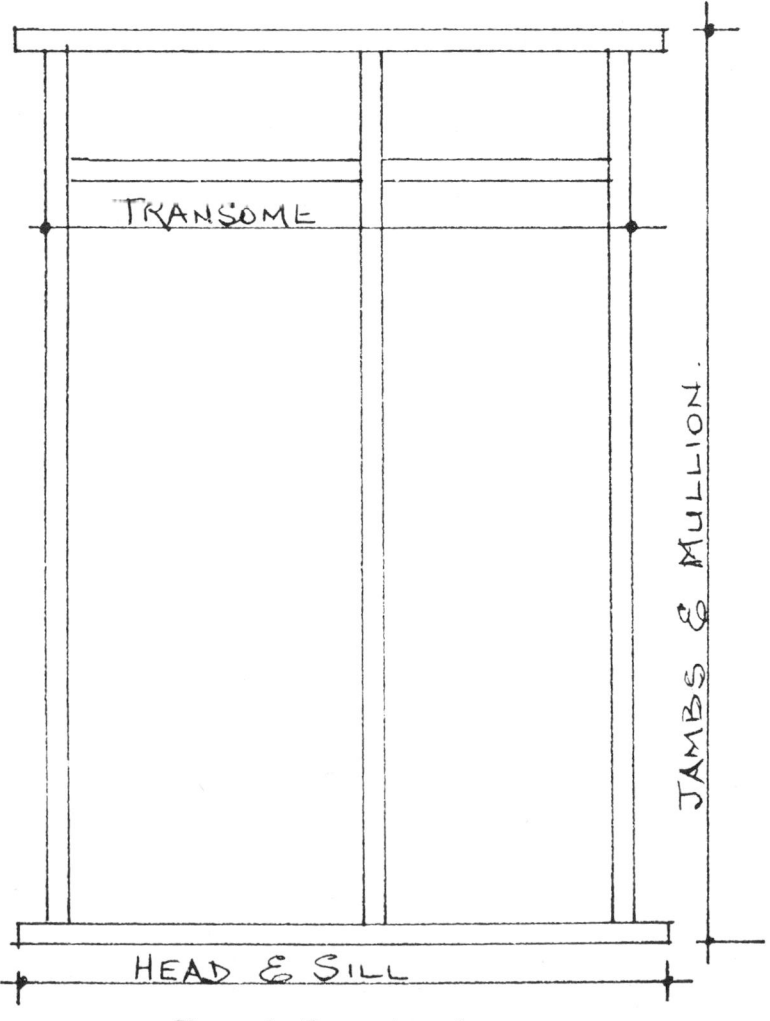

Figure 4   Frame lengths

Labours on frames and linings are given in the description of the items (SMM N17). (See Appendix B for Joinery Labours.)

The measurement of frames or linings is followed by the fixing items, e.g. dowels, cramps, fixing slips, bedding and pointing, cover fillets, architraves and the like.

PAINTING. Painting and decorating work is 'internal' or 'external' according to its position in the finished building (Practice Manual VI). Work carried out on members before they are fixed shall be so described stating if carried out on or off site (SMM V1).

Painting and decorating work shall be measured the area covered, and allowances made for the extra girth of edges, mouldings, panels, sinkings, corrugations and the like (SMM V2). The taker off is the sole judge as to what comprises an appropriate allowance, which may be made by measuring the work as a flat surface and multiplying the dimensions by an appropriate factor. Painting on panelled doors is therefore usually measured as a flat surface and multiplied by the appropriate factor, whereas the dimensions for flush doors are adjusted to include the extra girth of edges.

Painting shall be classified either as doors or as glazed doors (SMM V5).

Painting on frames, linings, and associated mouldings is grouped together (SMM V5).

OPENING ADJUSTMENT. Having completed the measurement of doors and frames, the taker off knows the exact size of the door openings and can then deduct the walls and finishings which have been measured over the opening.

Deductions of brickwork, facework or blockwork for lintels shall be measured as regards height to the extent only of full courses displaced (SMM G5, 14, 26). This is particularly important in the case of blockwork where the height of a lintel is frequently less than the height of a full course, so that no deduction is made. Furthermore no deduction is made for brickwork or blockwork displaced by the ends of lintels built into the wall (SMM G48).

Openings in cavity walls require a cavity gutter at the door head (SMM G37) and an item for closing the cavity at the jambs (SMM G9) although these are seldom indicated on the drawing.

Plate 9
FRAMED LEDGED AND BRACED DOOR

S.M.M. RULES   SECTION A.
                   F.1.3.4.18.21.
                   G.5.14.37.43.
                   N.1.17.19.20.31.32.
                   V.1.2.3.4.

APPROACH:–   Door   – Framed, ledged & braced door
                                 Ironmongery
                                 Glass – none
                                 Frame
                                 Fixing frame
                                 Painting.
              Opening
              adjustment – Walls & finishings
                                 Head inside & out
                                 Reveals inside & out
                                 Threshold inside & out.

(Door

1

West softwood matchboarded door size 800 x 2000 x 50 mm thk covered wi 25 mm thk t.g. + vee jtd matchbdg wi 25 mm thk ledges & braces 50 mm thk framing	**S.M.M. N1.17.19.** Particulars of quality given in Preamble in preference to description.

&

Pair 100 mm pressed steel butts scrd to sw.	**S.M.M. N32.** Usually specified in pairs. Nature of background – softwood.

&

Cylinder rim night latch Yale & Towne manuf. Co. Willenhall Ref 04 scrd to sw	**S.M.M. N32.** Appropriate reference from trade catalogue. Nature of background – softwood.

<u>The following in 1No. Door frame Set:-</u>		<u>S.M.M. N20.</u>

<u>S.M.M. N1.17.20</u>
Work measured net as fixed in position S.M.M. A3.
If frames are supplied with horns to be built into the work — the extra length of horns must be included in the length of each head.

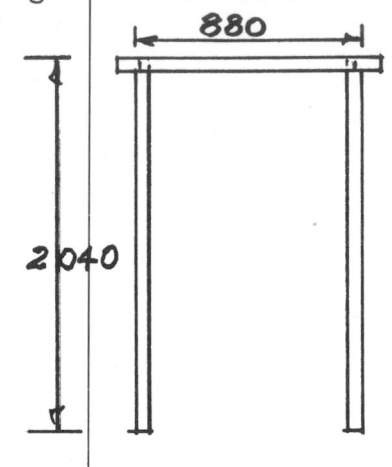

880

2040

	Jambs
	clear 2·000
	tenon 0·040
	<u>2·040</u>
2/2·04	100 × 50 mm wrot. S.w. reb & grooved jamb
	Head
	<u>clear 0·800</u>
	jambs 2/0·040 0·080
	<u>0·880</u>
0·88	100 × 50 mm Ditto head

<u>End - Door frame set</u>

<u>S.M.M. N31.</u>
In foot of each jamb — mortice in timber deemed to be included.

2	5 mm Dia mild steel dowel 100 mm long & mortice in precast conc step for ditto & run in neat ct.

<u>S.M.M. F9.</u>

<u>S.M.M. N31.</u>

2/ 4	25 × 3 mm galv m.s. fxg cramp 150 mm lgth one end turned up, holed & screwed to s.w. other end b.i. bkk

2/	2.04	Bed wood frame & plt b.s. in mastic	S.M.M. G43. Dimensions as for door frame set.
	0.88		

$1\frac{1}{8}$/	0.80	K.P.S. ③ wood doors &	S.M.M. V1.2.3.4. $1\frac{1}{8}$/ = allowance for edges and mouldings S.M.M. V2. Particulars of materials given in Preamble in preference to description.
	2.00	(new wk INTL	

Ditto (new wk EXTL

2/	2.04	Ditto wood frame n.e. 150 mm gth &	S.M.M. V5. Dimensions as for door frame set.
	0.88	(new wk INTL	

Ditto n.e. 150 mm gth
(new wk EXTL.

INTL.

EXTL.

(Opg adjust

door 0.800
jambs
2/0.040 ___0.080___
W = 0.880

Deduct overall size of door
and frame.

door 2.000
head ___0.040___
H = 2.040

	0.88	Ddt	S.M.M. G14.
	2.04	One bk wall fcgs o.s. fair face other side a.s.	

opg 0.880
B.I
2/0.100 ___0.200___
L = 1.080

S.M.M. F1.3.4.18.21.

1	Precast conc (1:2:4) lintol size 1080 × 225 × 150 mm fin fair on both faces & soffit reinf wi 2 No 12mm dia m.s. bars & bi. bkk in mor (1:1:6)	Particulars of materials given in Preamble in preference to description. Description quite adequate – bill diagram not required. S.M.M. F3. Classification – other concrete work.
0.88 0.15	Ddt One bk wall fcgs o.s. fair face other side ab.	S.M.M. G14. No deduction made for ends of lintol S.M.M. G48.
2/2.04	Fair return fair faced bkk H.B. wide & Fair return fcg bks HB wide	S.M.M. G14. Internal reveals S.M.M. G14. External reveals.
1	Precast conc (1:2:4) step size 1080 × 225 × 150 mm fin fair on tread & riser & bi. bkk in ct mor (1:3).	S.M.M. F1.3.4.18.21. Particulars of materials given in Preamble in preference to description. Description quite adequate – bill diagram not required. S.M.M. F3. Classification – other concrete work.
0.88 0.15	Ddt One bk wall in cb ab & Ddt E.O. cb for Tuscan fcgs ab.	S.M.M. G5. No deduction made for ends of steps S.M.M. G48. S.M.M. G14.
0.88	Ddt Hor DPC ab 225mm wide.	S.M.M. G37.

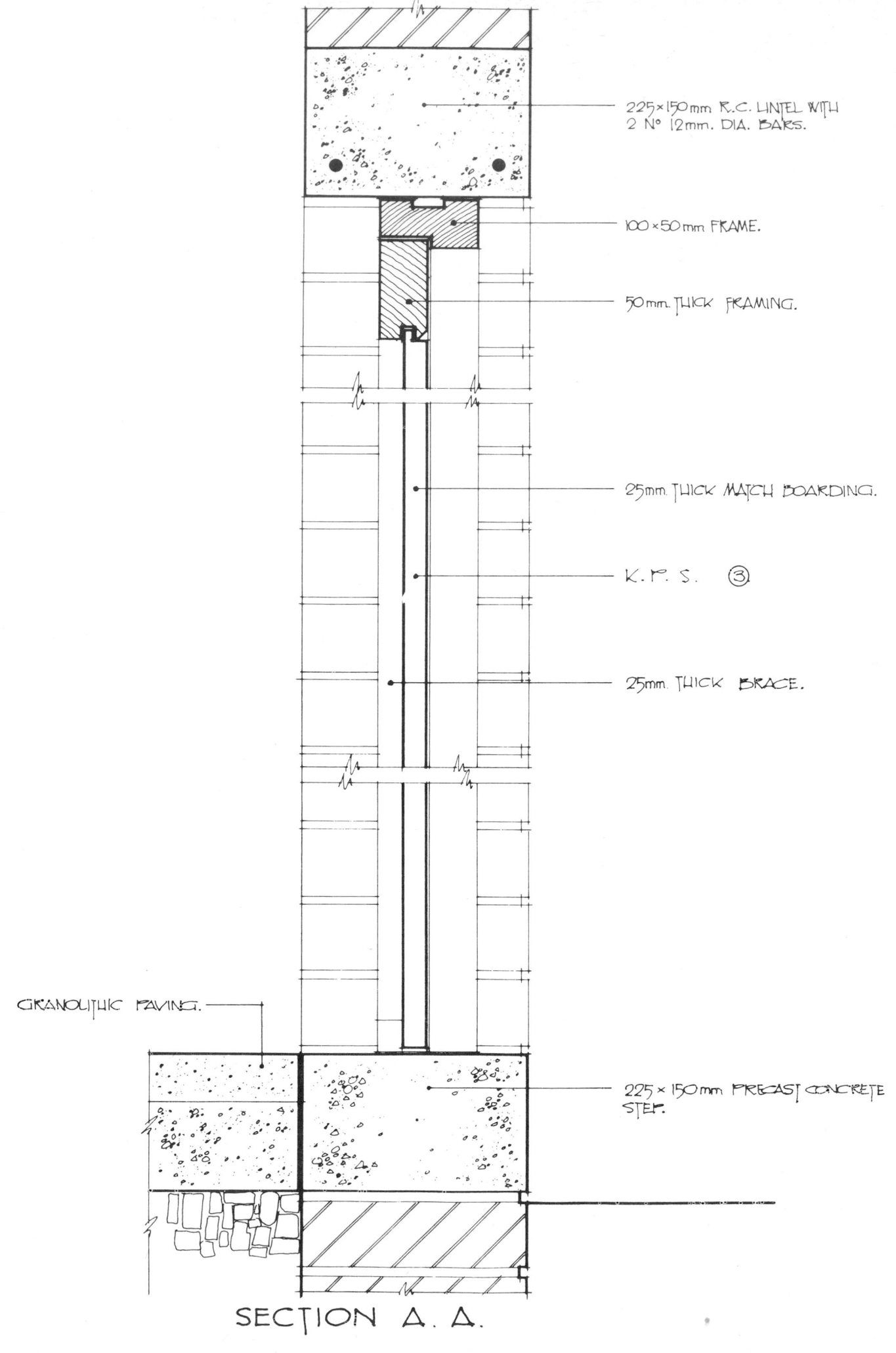

225×150mm R.C. LINTEL WITH 2 N° 12mm. DIA. BARS.

100×50mm FRAME.

50mm. THICK FRAMING.

25mm. THICK MATCH BOARDING.

K.P.S. ③

25mm. THICK BRACE.

GRANOLITHIC PAVING.

225 × 150mm PRECAST CONCRETE STEP.

SECTION A. A.

Plate 9
FRAMED LEDGED AND BRACED DOOR

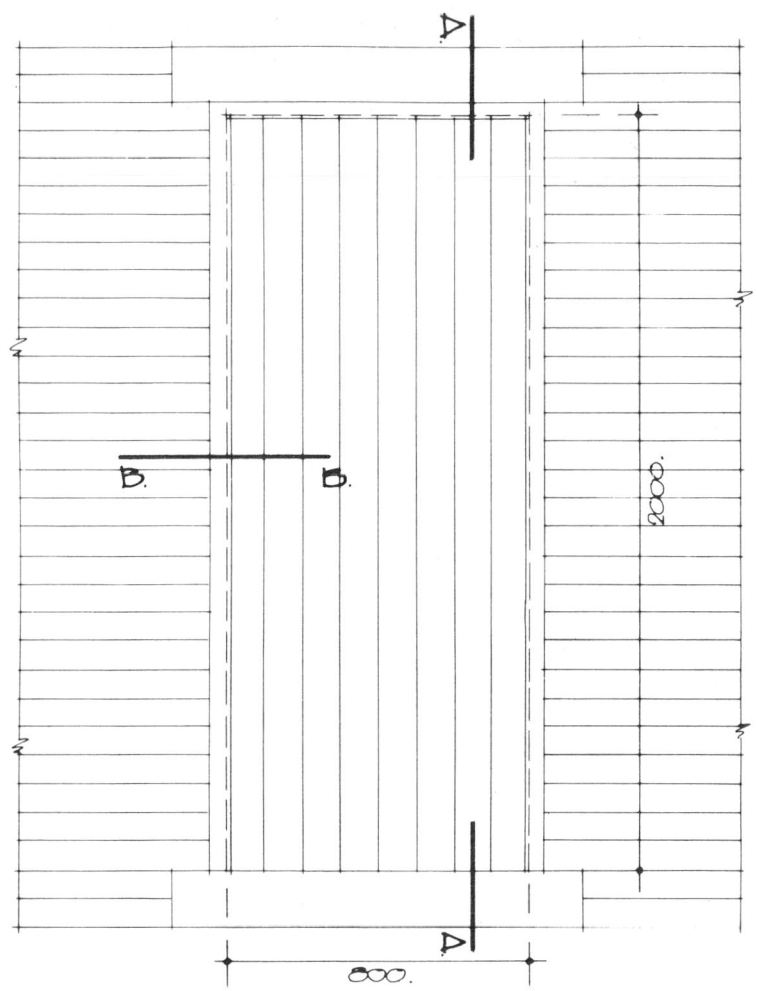

D

2000.

B. B.

D

800.

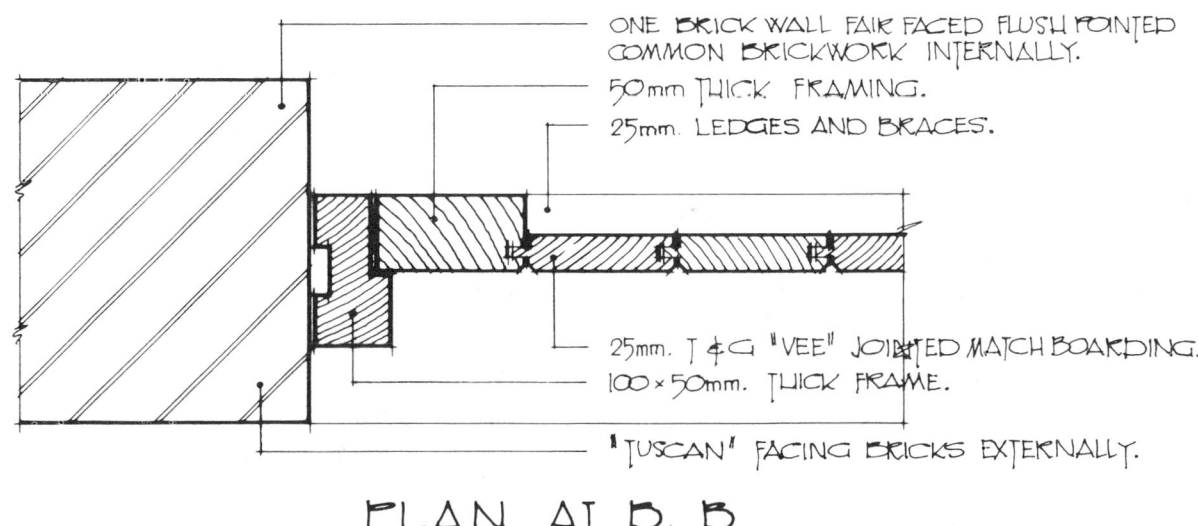

ONE BRICK WALL FAIR FACED FLUSH POINTED COMMON BRICKWORK INTERNALLY.

50mm. THICK FRAMING.

25mm. LEDGES AND BRACES.

25mm. T & G "VEE" JOINTED MATCH BOARDING.

100 x 50mm. THICK FRAME.

'TUSCAN' FACING BRICKS EXTERNALLY.

PLAN AT B. B.

IRONMONGERY.

100mm. PRESSED STEEL BUTTS.
CYLINDER RIM NIGHT LATCH. YALE AND TOWNE MANUFACTURING Cº LTD. REF. 04.
5mm. dia. MILD STEEL DOWELS 100mm LONG.

Plate 10
PANELLED DOOR

<u>S.M.M. RULES</u>	SECTION **A.** F.1.3.4.18.21. G.5.9.14.20.37. N.1.13.17.19.20.31.32. T.4.5.13.14. V.1.2.3.4.5.12.	

<u>APPROACH:-</u>	Door - Panelled door
	    Ironmongery
	    Glass - none
	    Frame
	    Fixing frame
	    Painting.
	Opening
	adjustment- Walls & finishings
	    Head inside & out
	    Reveals inside & out
	    Threshold inside & out.	

———————————

(Door

<u>1</u>	Wrot Iroko panelled door size 800 × 2000 × 50 mm thk wi 50 mm thk framg mo an solid both sides and 8 No 15 mm thk panels selected & kept clean for clear varnish	<u>S.M.M. N1.17.19.</u> Particulars of quality given in Preamble in preference to description
<u>1½</u>	Pair 100 mm pressed steel butts scrd to hdwd	<u>S.M.M. N32.</u> Usually specified in pairs. Nature of background - hardwood.

	1		Upright mortice lock Parker Winder Achurch Ltd., B'ham Ref SP 11901 wi set B.M.A. knob furniture Ref. SP 10111 scrd to hdwd	-	S.M.M. N32. appropriate references from trade catalogue.

$
B.M.A \; postal \; letter \; plate \; P.W.A. \; Ref. \; OP10682 \; ditto.
$

Following in 1No. Door frame set:-

S.M.M. N20.

S.M.M. N1.17.20.
Work measured net as fixed in position S.M.M. A3.
If frames are supplied with horns to be built into the work – the extra length of horns must be included in the length of each head.

Jambs.
door 2.000
head 0.040
2.040

2/2.04   100 x 50 mm Wrot iroko ad. reb & Grooved jamb

Head.
door 0.800
jambs
2/0.040 0.080
0.880

0.88   100 x 50 mm ditto head.

End. Door frame set:

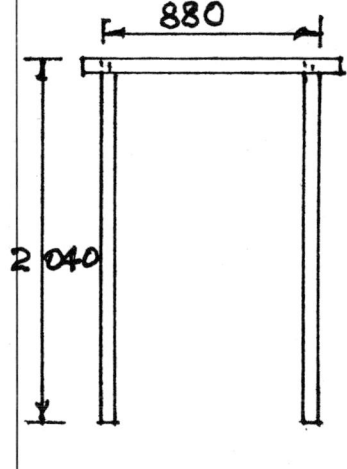

2	5 mm Dia Phosphor bronze dowel 100 mm long & mortice in precast conc steps for ditto & run in neat ct.	S.M.M. N31. In foot of each jamb. Mortice in timber deemed to be included. S.M.M. F9.
2/ 4	25 x 3 mm Phosphor bronze fixg cramp 150 mm gth one end turned up holed & scrd to cdwd other end b.i. bkk .	S.M.M. N31.
2/ 2.04 0.88	Bed wood frame & pt o.s. in mastic	S.M.M. G43. Dimensions as for door frame set.

Bedg
2/2.040   4.080
           0.880
          ‾‾‾‾‾‾
           4.960
less sktg
3/0.060   0.120
          ‾‾‾‾‾‾
           4.840

S.M.M. N13.

4.84	20 mm Wrot iroko cd. P. M.	

$1\frac{1}{8}$/ 0.80		
2.00		

Two cts clear
polyurethane
varnish wood
doors
&
(new wk
INTR

Ditto
(new wk
EXTR

S.M.M. V1.2.3.4.
Particulars of materials given in
Preamble in preference to description.

$1\frac{1}{8}$/ = allowance for edges and
mouldings S.M.M. V2.

2/ 2.04		
0.88		

Ditto wood frame
n.e. 150 mm gth
&
(new wk
INTR

Ditto
(new wk
EXTR.

S.M.M. V5.
Dimensions as for door frame set.

INTR.

Opg.
adjust

door 0.800
jambs
2/0.040 0.080
W = 0.880

door 2.000
head 0.040
H = 2.040

0.88		
2.04		

Ddt
HB. skin hollow
wall entirely
fcgs ab
&

Ddt
Forming cavities ab
&

Ddt
HB skin hollow wall
in cb ab.
&

Ddt
Two ct plaster ab walls

Deduct overall size of door and
frame.

S.M.M. G14.

S.M.M. G9.

S.M.M. G5.

S.M.M. T5.

		S.M.M. V12.

S.M.M. V12.

Not measured behind skirting.

2·040
Sktg 0·100
1·940

0·88
1·94

Ddt
Hang patterned
wall paper ab

S.M.M. F1.3.4.18.21.

qpg 0·880
BI
2/0·100 0·200
L= 1·080

Particulars of materials given in
Preamble in preference to description.

1

Precast conc (11214)
lintol size
1080 x 150 x 150 mm
all keyed for
plaster reinf wi
2 No 12 mm dia
m.s. bars & bi.
bkk in mor (1:1:6)

Description quite adequate —
bill diagram not required.

S.M.M. F3.
Classification — other concrete work.

0·88
0·15

Ddt
Forng cavities ab
&

S.M.M. G9.
No deduction made for ends of
lintol S.M.M. G48.

Ddt
H.B. skin hollow
wall in cb ab

S.M.M. G5.

lintel 1·080
O'hang
ends 2/0·100 0·200
1·280

S.M.M. G37.

BI edge 0·050
cavity 0·240
door head 0·080
0·370

86

1.28 0.37	One layer hessian based bituminous felt D.P.C. forming cavity gutter in hollow wall whg. 3.8 kg/m² bedd in mor (1:1:6) no allowce made for laps.		
2/ 1	Ends		
0.88	Brick-on-end flat arch entirely fcg bks ab. 225 mm wide on face 50 mm wide on exposed soffit in mor (1:1:6) ptd wi neat wps. jt	S.M.M. G14.20. Particulars as S.M.M. G20 but brickwork entirely facing bricks NOT E.O. facework.
0.88 0.23	~~Ddt.~~ H.B. skin hollow wall entirely fcg bks ab.	S.M.M. G14.
0.88 0.12 2/2.04 0.12	Two ct. plaster ab. walls n.e. 300 mm wide INTL	S.M.M. 'M.

87

0.88 / 2/2.04	External angle ditto	S.M.M. T4.	
0.88 / 0.12 / 2/1.94 / 0.12	Hang pattern wall paper ab.	S.M.M. V12. / Not measured behind skirting.	
	Opg 0.880 / Reveals 2/0.125 0.250 / NET 0.630 / Ddt:	S.M.M. N13.	
0.63	Ddt: / 20 x 100 mm / iroko sklg ab / &		
	Ddt: / 2 cts polyurethane / clear varnish / wood sklg / h.e. 150 mm gth / (new wk / INTL	S.M.M. V3.4.	
2/2.04	Closing cavities / 50 mm wide at / jambs of opgs wi / bkk incl one / layer hessian based / bit. felt vert. D.P.C. / whg 3.8 kg/m² / 112.5 mm wide / bedd in mor (1:1:6) / no allowce made / for laps / &	S.M.M. G9.37.	
	Fair return fcg / bks HB wide	S.M.M. G14.	

0.88 0.13		Terrazzo floor tiles ab & Ct & sand (1:3) level screeded bed for terrazzo ab	<u>S.M.M. T14.</u> <u>S.M.M. T13.</u>	
1		Precast conc (1:2:4) step size 1080 x 200 x 150 mm wi splayed riser & rodd nosg fin fair on tread & riser & bi. bkk in ct mor (1:3).	<u>S.M.M. F1.3.4.18.21.</u> Particulars of materials given in Preamble in preference to description. Description quite adequate – Bill diagram not required S.M.M. F3. Classification – other concrete work.	
0.88 0.15		Ddt. H.B. skin hollow wall entirely fgd ab & Ddt. Forng cavities ab	<u>S.M.M. G14.</u> <u>S.M.M. G9.</u>	
2/0.88		Ddt. Hoy D.P.C. ab 112.5 mm wide.	<u>S.M.M. G37.</u>	

89

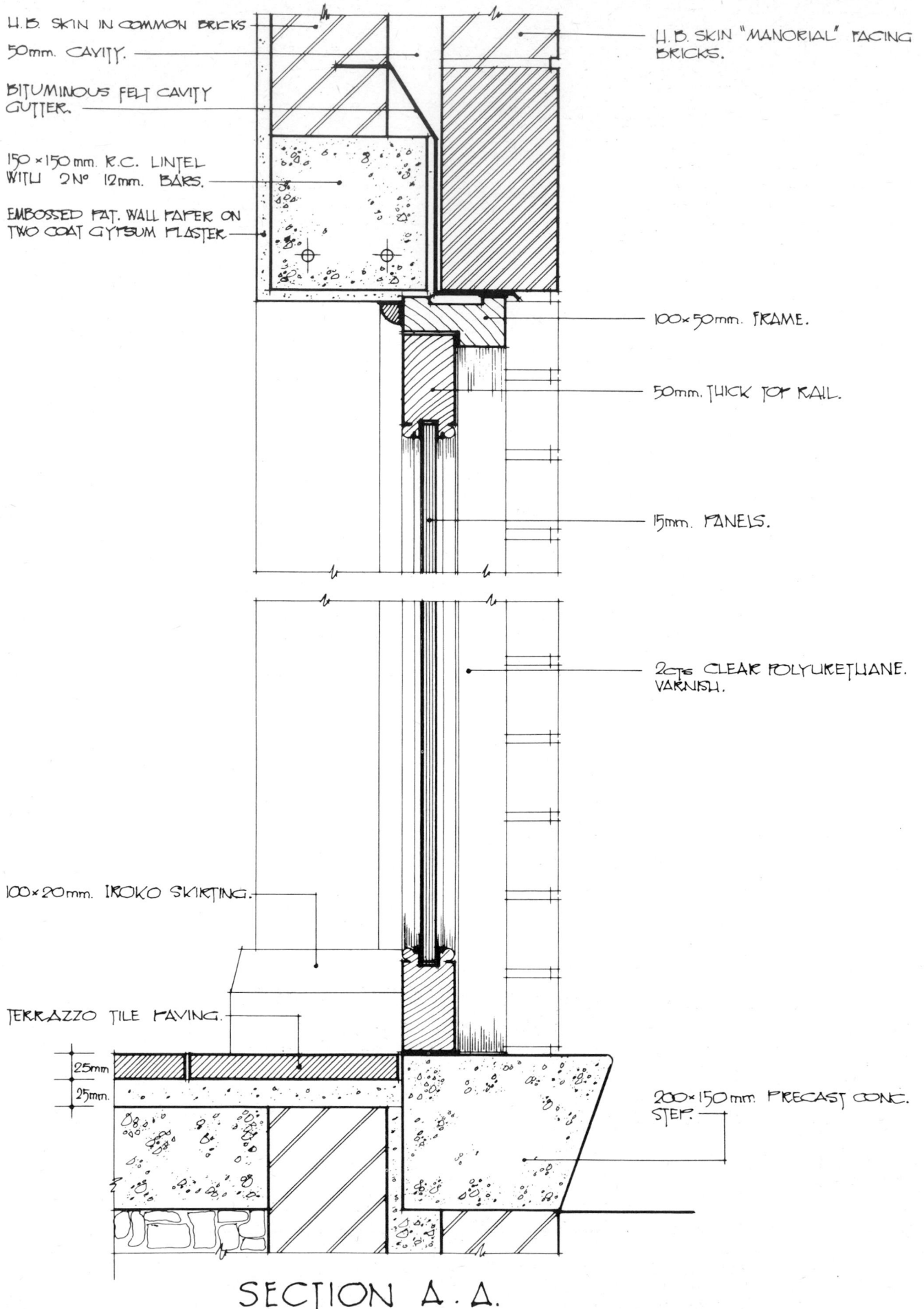

H.B. SKIN IN COMMON BRICKS
50mm. CAVITY.

BITUMINOUS FELT CAVITY
GUTTER.

150 × 150mm. R.C. LINTEL
WITH 2 N° 12mm. BARS.

EMBOSSED PAT. WALL PAPER ON
TWO COAT GYPSUM PLASTER.

H.B. SKIN "MANORIAL" FACING
BRICKS.

100 × 50mm. FRAME.

50mm. THICK TOP RAIL.

15mm. PANELS.

2 CTS CLEAR POLYURETHANE.
VARNISH.

100 × 20mm. IROKO SKIRTING.

TERRAZZO TILE PAVING.

25mm.
25mm.

200 × 150mm. PRECAST CONC.
STEP.

SECTION A·A.

Plate 10
PANELLED DOOR

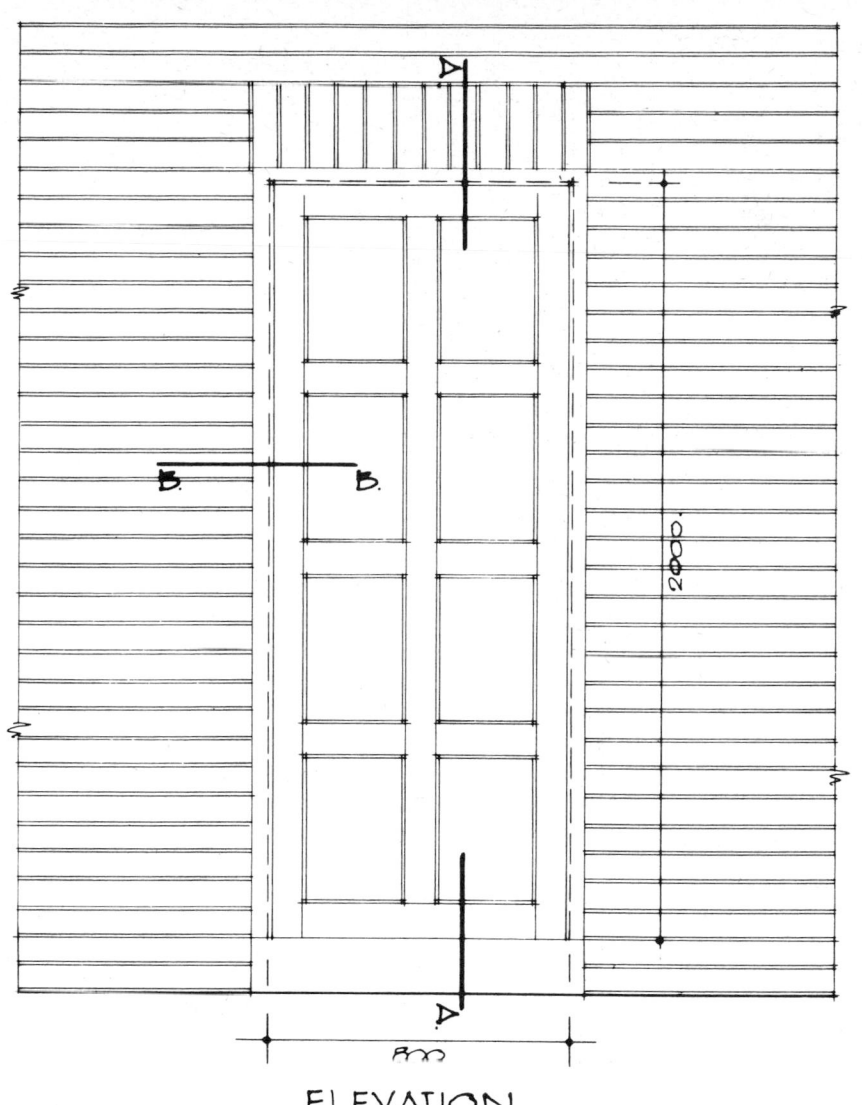

ELEVATION.

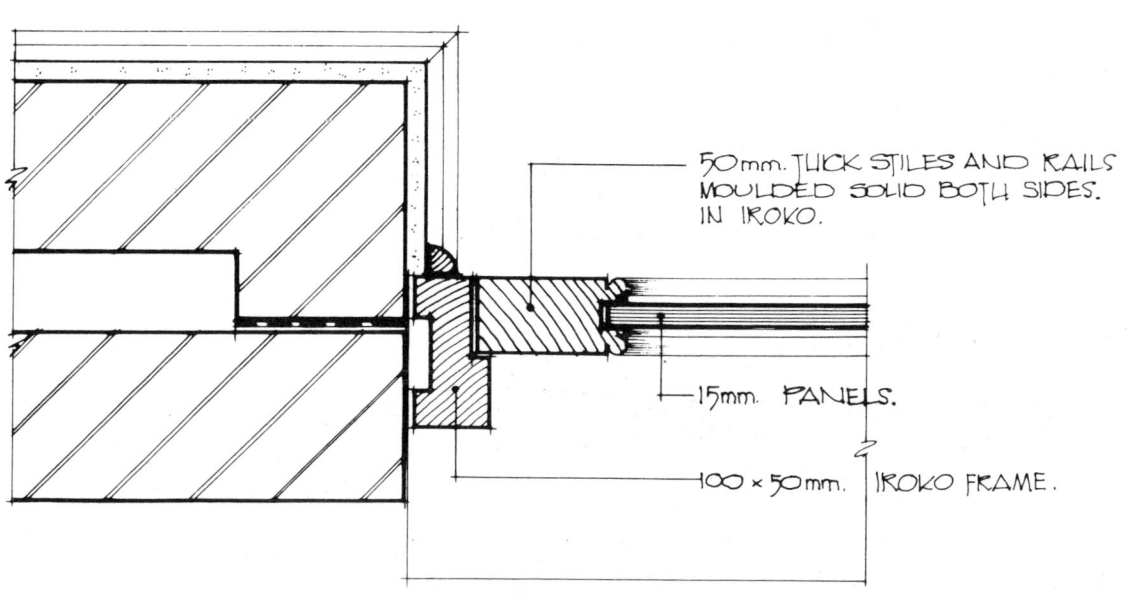

50mm. THICK STILES AND RAILS MOULDED SOLID BOTH SIDES. IN IROKO.

15mm. PANELS.

100 × 50mm. IROKO FRAME.

PLAN AT B. B.

IRONMONGERY :-

100mm PRESSED STEEL BUTTS.
UPRIGHT MORTICE LOCK P.W.A. REF. SF 11901
B.M.A KNOB FURNITURE " " " 10111.
B.M.A. LETTER PLATE " " OF 10682.

Plate 11
FLUSH DOOR

S.M.M. RULES SECTION A.
     F.1.3.4.18.21.
     G.26.27.
     N.1.4.11.13.17.19.20.32.
     T.5.
     V.1.2.3.4.5.

APPROACH:- Door - Flush door
        Ironmongery
        Glass - none
        Frame
        Fixing frame
        Painting
    Opening
    adjustment- Walls and finishings
        Head inside and out
        Reveals inside and out
        Threshold inside and out.

————————

(Door

1	Block pattern flush door J. Hills & Sons Ltd. Stockton-on-Tees "CLYMAX" size 826 X 2040 X 44 mm thk.	S.M.M. N1.17.19. A6.
	&	
	Pair 100 mm pressed steel butts scr to Sw.	S.M.M. N32. Usually specified in pairs. Nature of background - softwood.
	&	
	Tubular mortice latch Yale & Town manuf. Co Willenhall REF M 888 scr to Sw	S.M.M. N32. - Appropriate reference from manufacturers catalogue.

92

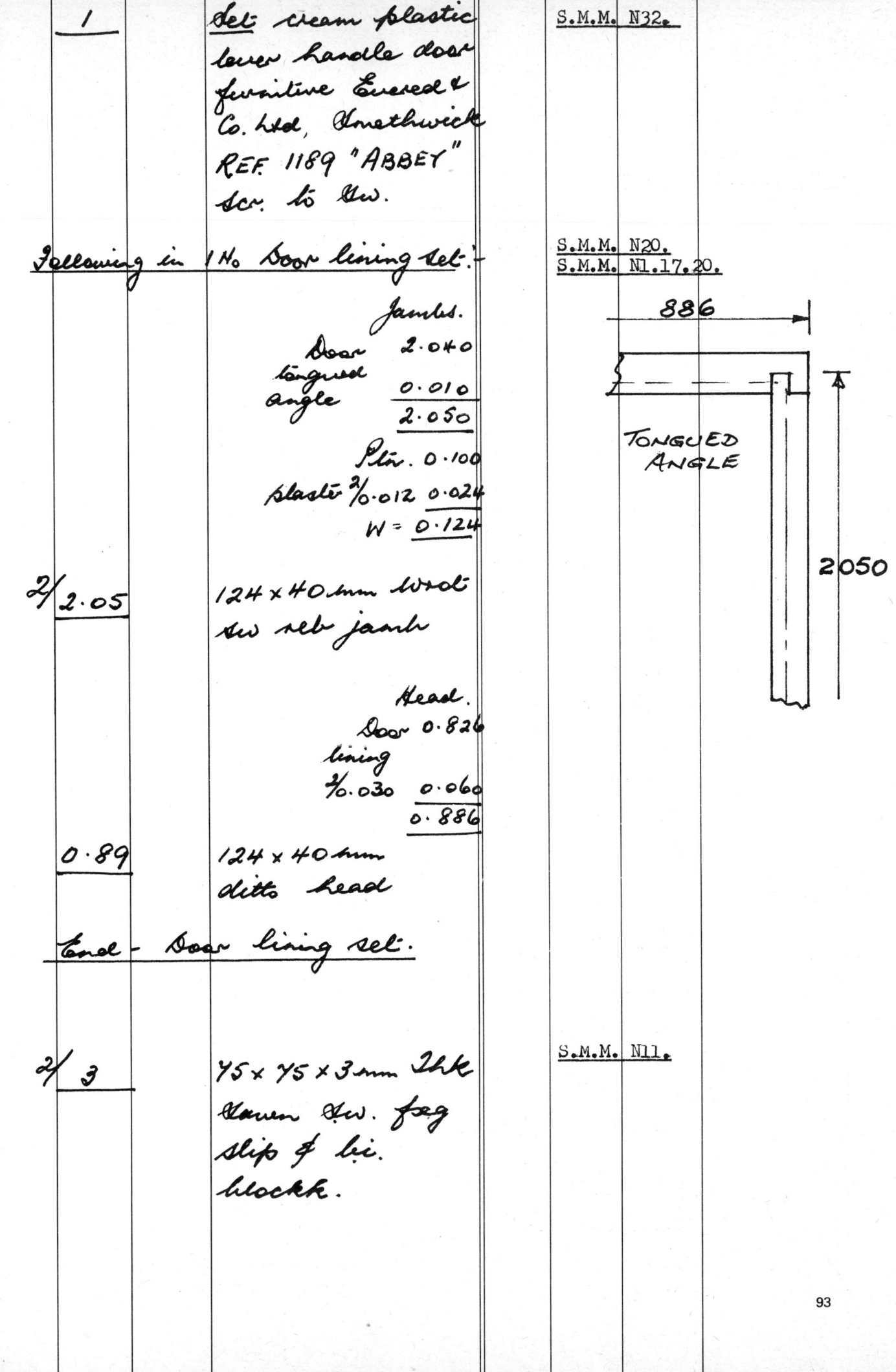

1	Set cream plastic lever handle door furniture Evered & Co. Ltd, Smethwick REF. 1189 "ABBEY" Scr. to sw.	S.M.M. N32.

Following in 1 No. Door lining set:-

S.M.M. N20.
S.M.M. N1.17.20.

Jambs.

Door tongued angle	2.040
	0.010
	2.050

Ptn. 0.100

plaster ²/0.012 0.024

W = 0.124

886

TONGUED ANGLE

2050

| 2/2.05 | 124 × 40 mm wrot sw reb jamb |

Head.

Door lining 0.826

²/0.030 0.060

0.886

| 0.89 | 124 × 40 mm ditto head |

End - Door lining set.

| 2/3 | 75 × 75 × 3 mm thk galven. sw. fixg slip & bit. blockk. | S.M.M. N11. |

Door 0.826
lining &
arch. 2/0.090 0.180
              1.006

Ht.
Door 2.040
lining
& arch 0.090
    2/ 2.130 = 4.260
            5.266

2/5.24

75 x 25 mm Wrot
dw splayed arch.

S.M.M. V1.2.3.5.
Particulars of material given in
Preamble in preference to description.

Door 0.826
edge 0.044
    0.870

Door 2.040
½/ lap 0.022
    2.062

Door dimensions adjusted to include
extra girth of edges. S.M.M.V2.

2/0.84
2.06

K.P.S. ②
wood door
   (new wk
   Intl.

arch.
2/0.025 0.050
2/0.075 0.150
4/0.015 0.030
lining — 0.020
        0.044
        0.010
        0.080
        0.030
        0.414

S.M.M. V5.

= 414 mm

Dimensions as for door lining.

2/2.05
0.41

0.89
0.41

Ditto wood
lining & assoc.
mouldings.
   (new wk
   (Intl

		(Opg (adjust.	

	Door	0.826
	lining 2/0.030	0.060
	W =	0.886

	Door	2.040
	lining	0.030
	Ht	2.070

S.M.M. G26.27.
Deduct overall size of door and lining.

0.89	_Ddt._
2.07	100 mm thk conc block ptn ab

2/0.89	_Ddt._
2.07	Two ct plaster a.b. walls

S.M.M. T5.

S.M.M. V4.

Painting not measured behind skirting.

	Ht.	2.070
	Sktg	0.100
		1.970

2/0.89	_Ddt._
1.97	2ce emulsion ditto

S.M.M. F1.3.4.18.21.
Particulars of materials given in Preamble in preference to description.

	Opg	0.886
	B.I. 2/0.100	0.200
	L =	1.086

1	Precast conc (1:2:4) lintol size 1086 x 100 x 150 mm All keyed for plaster reinf wi 1 No 12 mm Ø m.s. bar & bic. blockk in mor (1:2:9).

95

No deduction made on partition.
Full block course not displaced.
Block course 225mm high, lintol
150mm high.   S.M.M. G26.

2/1.01	*Ddt:* 25 × 100 mm &w. sklg ab &	S.M.M. N13. Dimension from waste calculation for architrave.
	Ddt: K. P. S. ② wood sklg n.e. 150 mm gth (new work Intl	S.M.M. V4.
0.89 0.10	25 mm thk wrot &w t & g. board floorg ab.	S.M.M. N4.

96

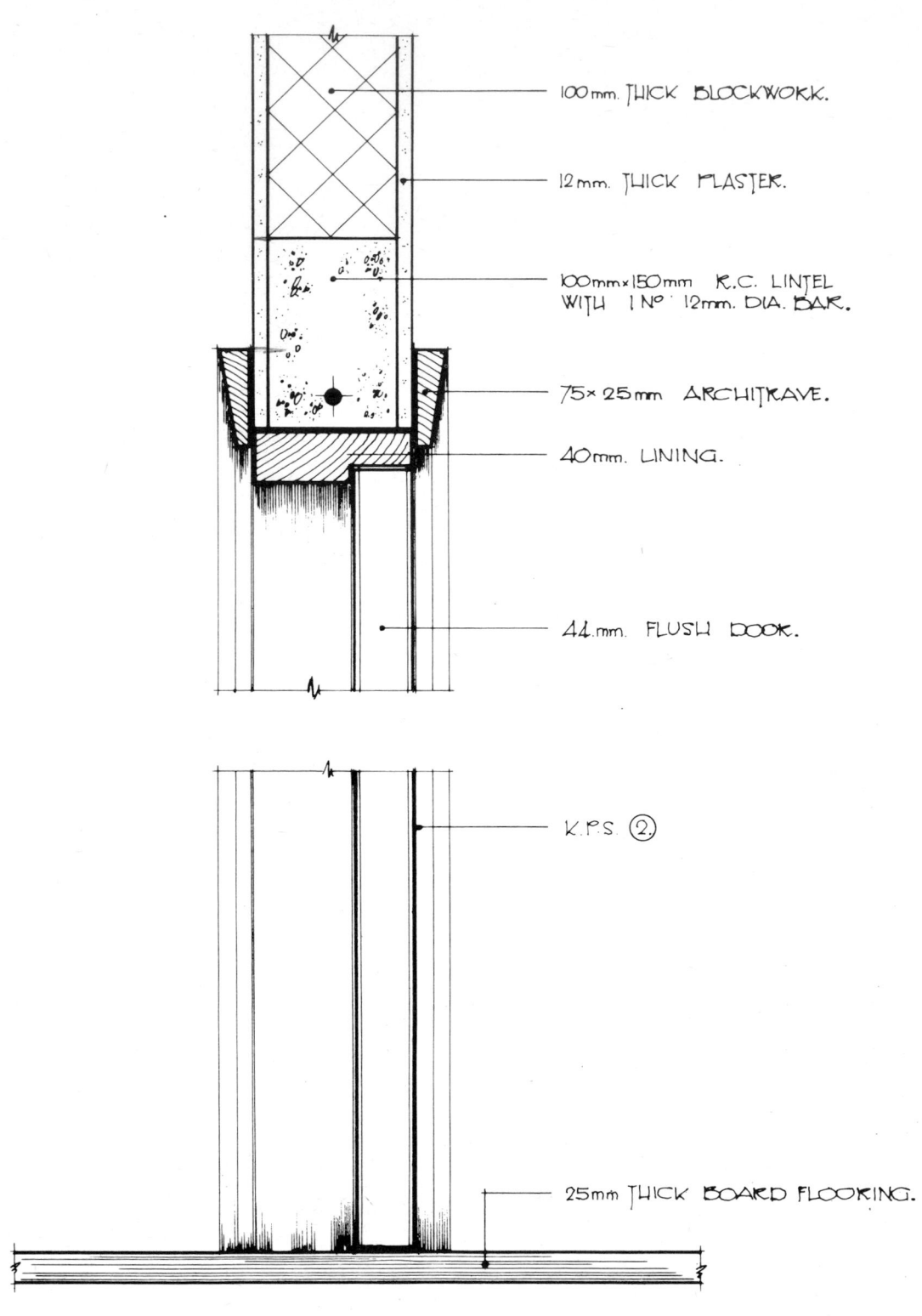

100mm. THICK BLOCKWORK.

12mm. THICK PLASTER.

100mm×150mm R.C. LINTEL
WITH 1 No. 12mm. DIA. BAR.

75×25mm ARCHITRAVE.

40mm. LINING.

44.mm. FLUSH DOOR.

K.P.S. ②

25mm THICK BOARD FLOORING.

SECTION A. A.

Plate 11
FLUSH DOOR

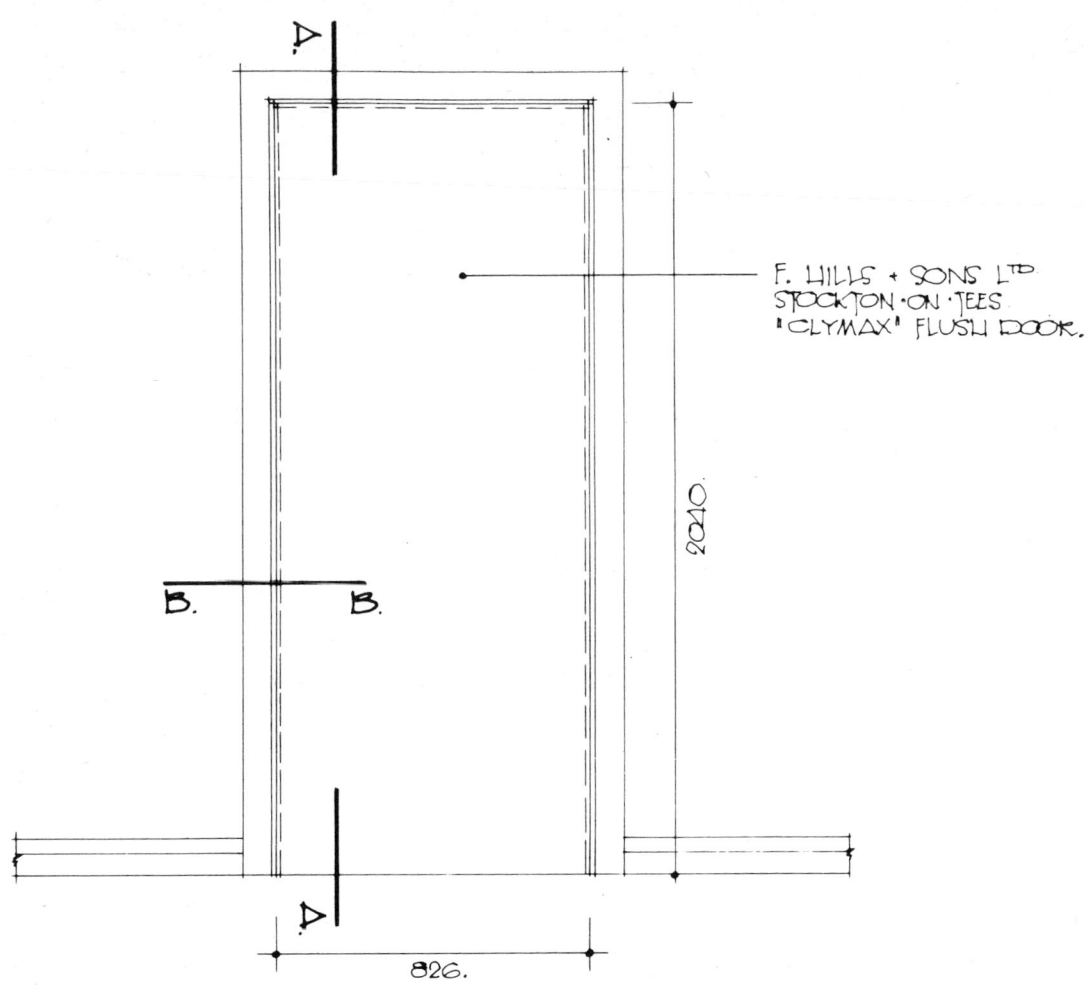

F. HILLS + SONS LTD.
STOCKTON·ON·TEES.
"CLYMAX" FLUSH DOOR.

2040.

826.

ELEVATION.

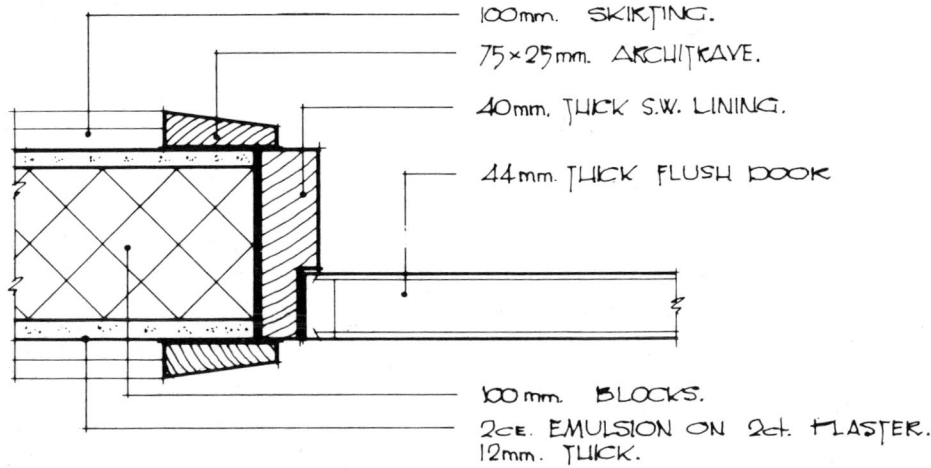

100mm. SKIRTING.

75×25mm. ARCHITRAVE.

40mm. THICK S.W. LINING.

44mm. THICK FLUSH DOOR

100mm. BLOCKS.

2cs. EMULSION ON 2ct. PLASTER.
12mm. THICK.

PLAN AT B. B.

IRONMONERY :-

100mm. PRESSED STEEL BUTTS.
TUBULAR MORTICE LATCH. YALE AND TOWNE MANUFACTURING Cº REF. M 888.
CREAM PLASTIC LEVER HANDLE FURNITURE. EVERED AND Cº REF. 1188 " ABBEY."

Plate 12
GLAZED DOOR AND SIDELIGHT

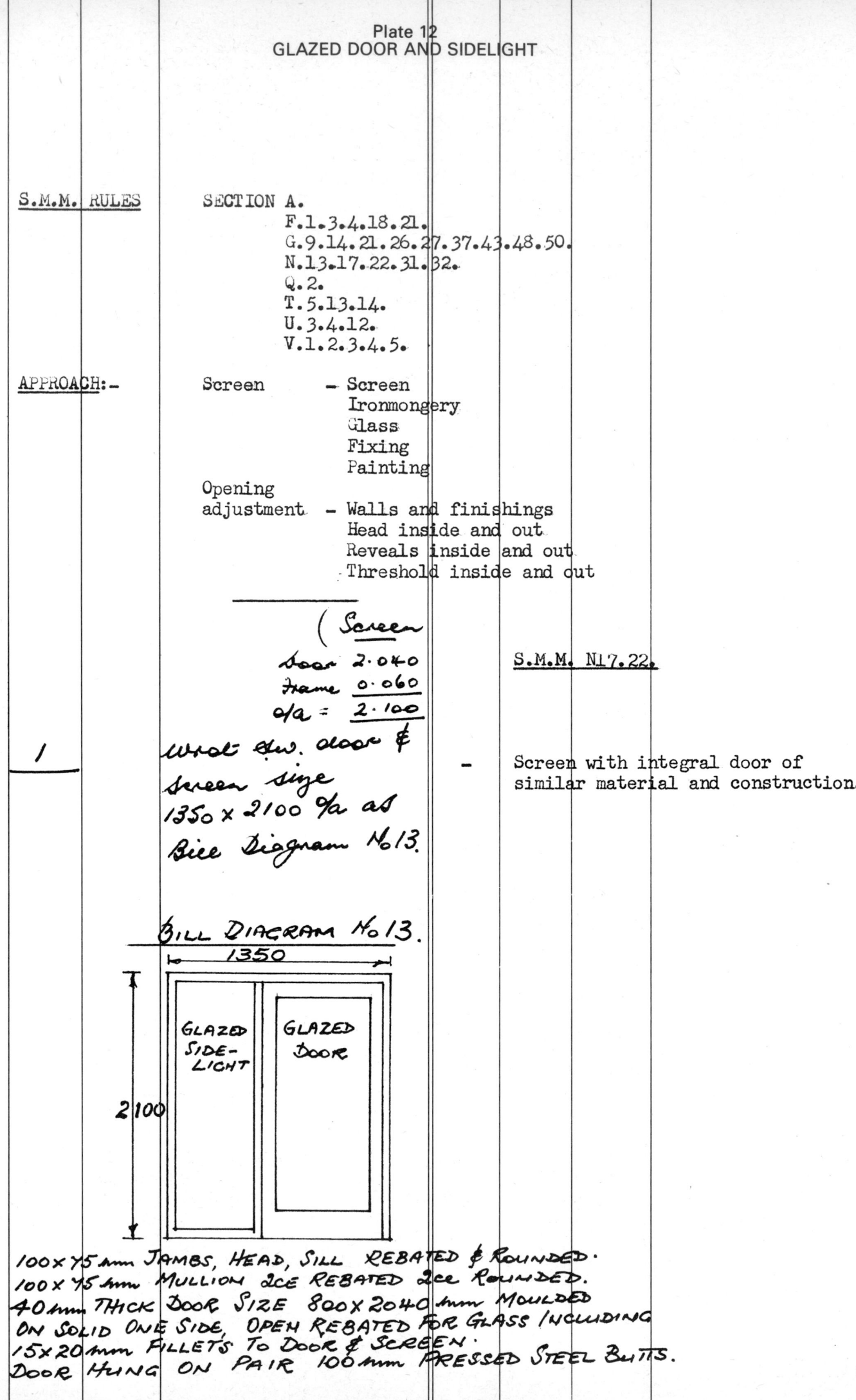

S.M.M. RULES   SECTION A.
                      F.1.3.4.18.21.
                      G.9.14.21.26.27.37.43.48.50.
                      N.13.17.22.31.32.
                      Q.2.
                      T.5.13.14.
                      U.3.4.12.
                      V.1.2.3.4.5.

APPROACH:-   Screen   - Screen
                                      Ironmongery
                                      Glass
                                      Fixing
                                      Painting
                     Opening
                     adjustment   - Walls and finishings
                                      Head inside and out
                                      Reveals inside and out
                                      Threshold inside and out

(Screen

door 2·040   S.M.M. N17.22.
frame 0·060
o/a = 2·100

1   Wrot. sw. door &   - Screen with integral door of
         screen size   similar material and construction.
         1350 × 2100 o/a as
         Bill Diagram No 13.

BILL DIAGRAM No 13.

|← 1350 →|

GLAZED SIDE-LIGHT   GLAZED DOOR

2100

100×75mm JAMBS, HEAD, SILL REBATED & ROUNDED.
100×75mm MULLION 2ce REBATED 2ce ROUNDED.
40mm THICK DOOR SIZE 800×2040mm MOULDED
ON SOLID ONE SIDE, OPEN REBATED FOR GLASS INCLUDING
15×20mm FILLETS TO DOOR & SCREEN.
DOOR HUNG ON PAIR 100mm PRESSED STEEL BUTTS.

1	Cylinder rim night latch P.C. £5.00 ser to fix.	S.M.M. N32. Prime Cost Sum included where manufacturers reference not specified.	

&

Satin anodised aluminium letter plate P.C. £5.00 ditto

S.M.M. N32.

glass

Door 0.800

Stile 0.100
rebate 0.015
2/ 0.085   0.170
     W = 0.630

Door 2.040

Top rail
       0.100
rebate 0.015 0.085

Bot rail
       0.200
rebate 0.015 0.185 0.270
            Ht = 1.770

0.630 × 1.770 = 1.115 m²   —   classification over 1.00m²

S.M.M. U3.4.

0.63	5 mm Tht Spotlyte
1.77	glass & glaze to wood wi P.V.C. strip & wood fillets fxd wi brads in panes over 1.00 m²

2/0.63	Bed edge panes	S.M.M. U12.
2/1.77	in P.V.C. strip	Glazing fillets included with door and screen. S.M.M. N13.17.

$^{o}/_{a}$ 1.350

Door   0.800

frame

2/0.060   0.120

mullion   0.045   0.965
                    —————————
        W = 0.385

Door   2.040

frame   0.060

Ht = 1.980

$0.385 \times 1.980 = 0.762 \, m^2$ —

S.M.M. U3.4.

classification $0.50 - 1.00 m^2$

0.39
1.98
————

5mm thk Spotlyte glass a.b. in panes $0.50 - 1.00 \, m^2$

2/0.39
2/1.98
————

Bed edge glass in P.V.C. strip

S.M.M. U12.

(Jog

2
————

5mm ∅ mild steel dowel 100 mm long

&

mortice in bkk for ditto & run wi int cl.

S.M.M. N31.
To door jambs only.

S.M.M. G50.

2/ 4
————

25 × 3 mm Galv M.S. peg cramps 150 mm gth one end turned up holed & scr. to fw. other end bi. bkk

S.M.M. N31.

%a   1.350
Door   0.800
                0.550

1.35
0.55
2/2.10

Bed wood
frame & pt one
side in mastic

              2.100
Sklg   0.100
              2.000

1.35
2/2.00

15 mm wood
& quadrant
mould.

              0.762
              1.115
        2) 1.877
Avg 0.939

1.35
2.10

K. P. S. ③ glazed
wood door &
screen avg
large panes

&   (new wk
          Intl

Ditto   (new wk
          Extl

          (Opg
          adj

Ddt
H.B. skin hollow
wall entirely fegs
ab

1.35
2.10

&

Ddt
Facing cavity ab

103

1.35 2.10	~~Ddt~~ 100 mm thk conc block skin hollow wall ab &	S.M.M. G26.27.
	~~Ddt~~ Two ct plaster ab. walls	S.M.M. T5.
	Ht 2·100 Skg 0·100 2·000	S.M.M. V4. Paint not measured behind skirting.
1.35 2.00	~~Ddt~~ 2cs emulsion ditto	
	$\frac{dg}{2/0·100}$ $\begin{array}{r}1·350\\0·200\\\hline 1·550\end{array}$ BI	S.M.M. F1.3.4.18.21.
1	Precast conc (1:2:4) splayed lintol size 1550 × 150 × 225 mm all keyed for plaster reinf wi 2 No 12 mm ∅ m.s. bars & b.i. blockk in mor (1:1:6).	Particulars of materials given in Preamble in preference to description Description quite adequate – bill diagram not required. S.M.M. F3. Classification – other concrete work.
1.35 0·23	~~Ddt~~ 100 mm thk conc block skin hollow wall ab.	S.M.M. G26.27.48. Full block course displaced by lintol. No deduction made for ends of lintol – S.M.M. G48.

	1	Galv. steel standard lintel 1550 mm long 229 mm dp. whg 9.4 kg/m Dorman Long (Steel) Ltd., Middlesbro' & bi. blockk in mor (1:1:6)	S.M.M. Q2. A6. No specific provision made for this item in S.M.M. – clause Q2 most appropriate.	

	1.35			
	0.08	Two cts plaster ab walls n.e. 300 mm wide (INTR	S.M.M. T5.	
2/	2.10			
	0.08	&		
		2ce Emulsion ditto	S.M.M. V4.	

| | 1.35 | Extl angle plaster | S.M.M. T5. | |
| 2/ | 2.10 | | | |

Ddt: opg 1.350
Add reveals
2/0.100 0.200
Ddt: NET 1.150

| | 1.15 | 20 x 100 mm sw sklg ab & | S.M.M. N13. | |

Ddt:
K.P.S. (2) wood sklg
n.e. 150 mm gth.
(new int
INTR.

| 2/ | 2.10 | Closing cavities 50 mm wide at jambs of opg in blockk incl one layer hessian based bit. felt vert D.P.C. whg 3.8 kg/m² 100 mm wide bedd in mor (1:1:6) no allowce made for laps. | S.M.M. V3.4.

S.M.M. G9.37. | |

105

2/ 2.10	Fair return feg like H.B. wide		S.M.M. G14. External reveals.
1.35 0.11	P.V.C. floor tiles a.b. &		S.M.M. T14.
	Ct & sand (1:3) 35 mm thk level trowelled bed a.b		S.M.M. T13.
1.35	Hoz threshold entirely feg like a.b. 160 mm on bed 112.5 mm high formed all bullnosed headers on edge bedd & ptd in ct mor (1:3).		S.M.M. G21.
2	Ends ditto		
1.35 0.08	Ddt. H.B. skin hollow wall entirely fegs ab		S.M.M. G14.
1.35	Ddt. Hoz D.P.C. 112.5 mm wide a.b.		S.M.M. G37.

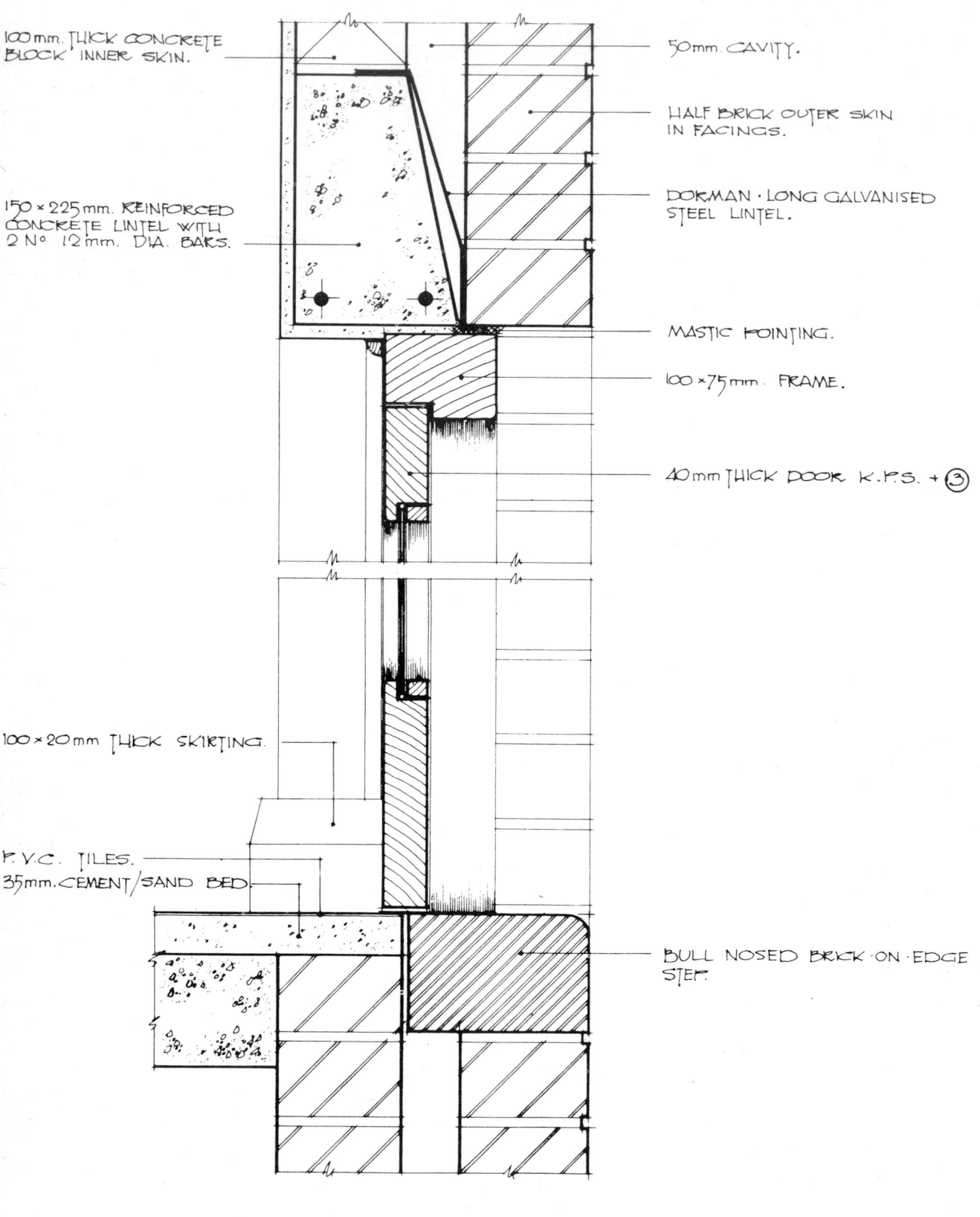

100mm. THICK CONCRETE BLOCK INNER SKIN.

150 × 225mm. REINFORCED CONCRETE LINTEL WITH 2 N° 12mm. DIA. BARS.

100 × 20mm THICK SKIRTING.

P. V. C. TILES.
35mm. CEMENT/SAND BED.

50mm. CAVITY.

HALF BRICK OUTER SKIN IN FACINGS.

DORMAN · LONG GALVANISED STEEL LINTEL.

MASTIC POINTING.

100 × 75mm. FRAME.

40mm THICK DOOR K.P.S. + ③

BULL NOSED BRICK · ON · EDGE STEP.

SECTION A. A.   SCALE 1 : 5.

Plate 12
GLAZED DOOR AND SIDELIGHT

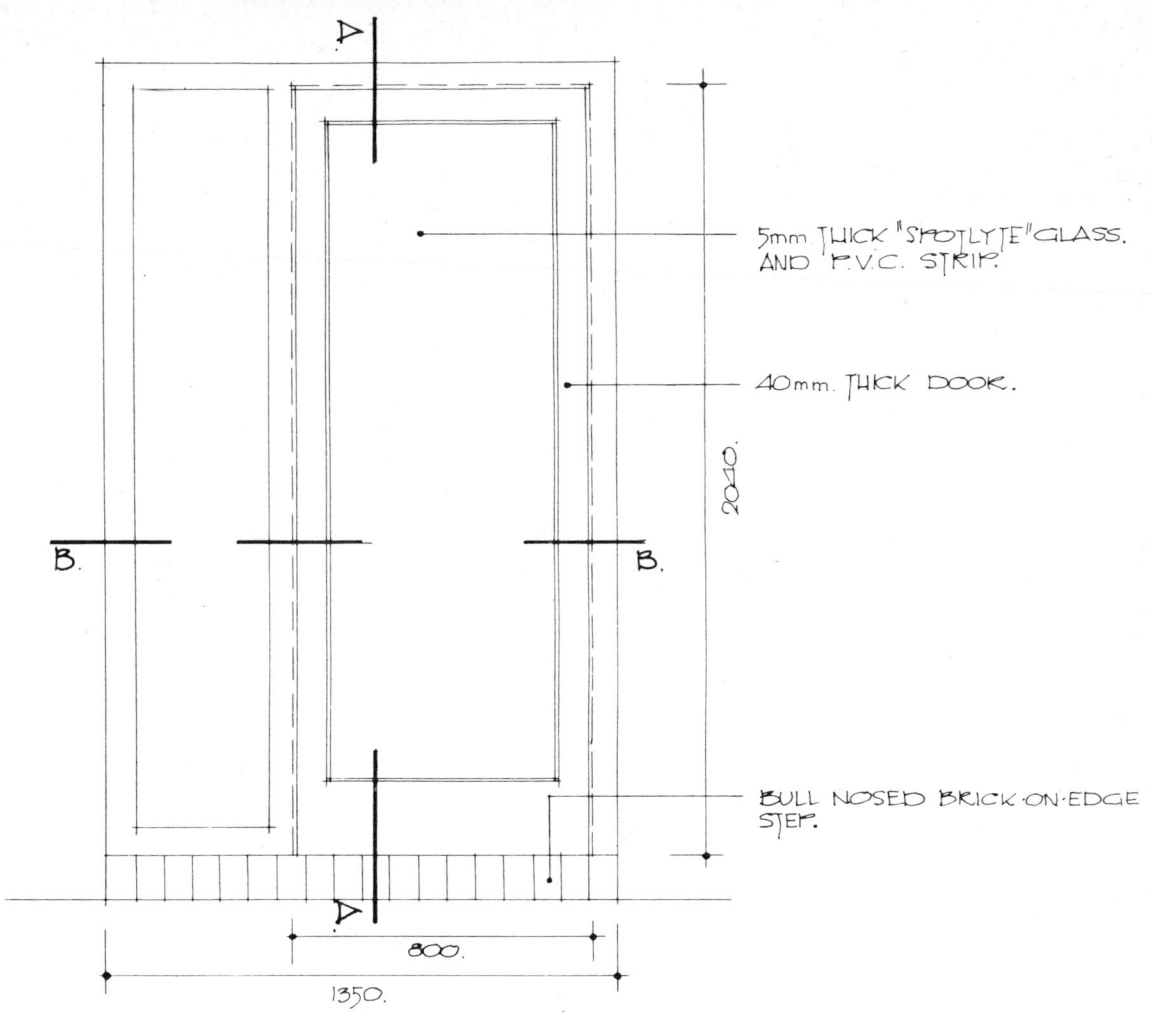

5mm. THICK "SPOTLYTE" GLASS.
AND P.V.C. STRIP.

40mm. THICK DOOR.

2010.

BULL NOSED BRICK·ON·EDGE
STEP.

800.

1350.

ELEVATION.   SCALE. 1 : 20.

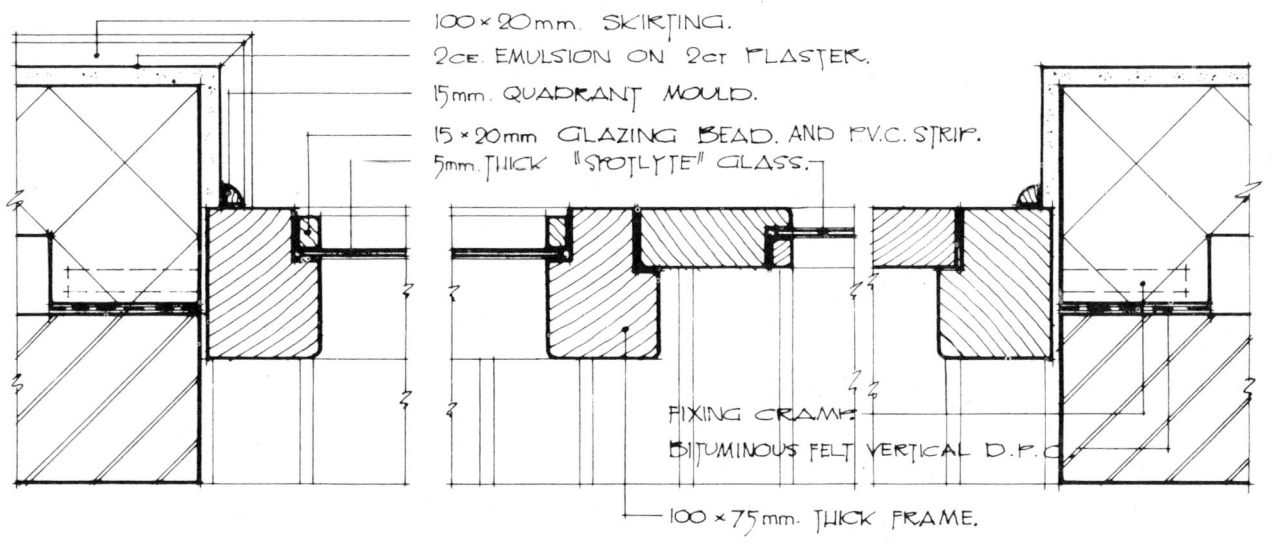

100 × 20mm. SKIRTING.
2 CE. EMULSION ON 2 CT PLASTER.
15mm. QUADRANT MOULD.
15 × 20mm GLAZING BEAD. AND P.V.C. STRIP.
5mm. THICK "SPOTLYTE" GLASS.

FIXING CRAMP.
BITUMINOUS FELT VERTICAL D.P.C.

100 × 75mm. THICK FRAME.

PLAN AT B.B.   SCALE 1 : 5.

IRONMONGERY :-
100mm PRESSED STEEL BUTTS.
CYLINDER RIM NIGHT LATCH   P.C. £5.00
S.A.A. LETTER PLATE.   P.C. £5.00

Plate 13
FLUSH DOORS 'A'

S.M.M. RULES   SECTION A.
             F.1.3.4.18.21.
             G.26.27.
             N.1.4.13.17.19.20.31.32.
             T.5.
             U.3.4.12.
             V.1.2.3.4.5.12.

APPROACH:-   Doors   - Flush doors
                                Ironmongery
                                Glass
                                Frame
                                Fixing frame
                                Painting
            Opening
            adjustment - Walls and finishings
                                Head inside and out
                                Reveals inside and out
                                Threshold inside and out.

(Doors

     W opg   1.670
frame $\frac{2}{0.040}$ 0.080
              2)1.590
                0.795
     rebate 0.005

Extreme W = 0.800

     Ht opg 2.040
     frame   0.040
       Ht   2.000

S.M.M. N1.17.19.

Particulars of quality given in
Preambles in preference to
description.

2/ 1   Flush door size
800 x 2000 x 40 mm thk
wi skeleton core
faced b.s. wi 3mm
thk Sapele
veneered hardboard
fxd wi synthetic
adhesive incl solid
lippg to all four edges
wi rebated meeting
stile & opg for glass
size 250 x 1000 mm incl
18 x 20 mm Sapele glazg
fillets b.s.

2/ 1	Pair 100 mm pressed steel butts scr hdwd	S.M.M. N32. Usually specified in pairs. Nature of background – hardwood.
1	Related mortice latch wi set lever handle furniture P.C. £5.00 to hdwd	S.M.M. N32.
		Prime Cost Sum included where manufacturers reference not specified.
	& 100 mm Barrel bolt P.C. £2.00 ditto	S.M.M. N32.
	0.250 × 1.000 = 0.25m² –	S.M.M. U3.4. classification 0.10 – 0.50m²
2/ 0.25 1.00	5 mm thk clear sheet glass & glazg to wood wi P.V.C. strip & wood fillets fixed wi brass cups & scrs in panes 0.10 – 0.50 m²	
2/2/ 0.25 2/2/ 1.00	Bed edge panes in P.V.C. strip	S.M.M. U12.
		Glazing fillets included with door S.M.M. N13.

Following in 1No. Door frame set:-

2/2.04   100 x 50 mm Wrot
         Sapele rebated
         jamb selected &
         kept clean for
         clear varnish

1.67   100 x 50 mm
         ditto head

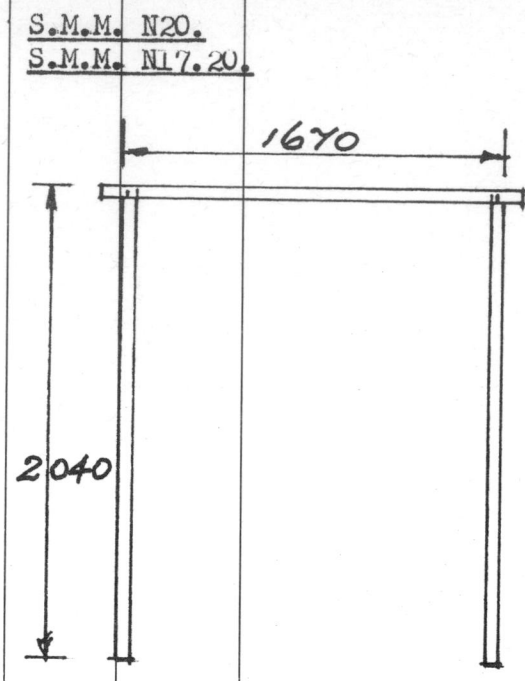

1670

2040

Work measured net as fixed in
position S.M.M. A3. If frames
are supplied with horns to be
built into the work — the extra
length of horns must be included
in the length of each head.

S.M.M. N31.

End - Door frame set.

2/4   25 x 3 mm Galv
         m.s. jeg cramp
         150 mm gth one
         end turned up
         holed & scr to
         hdwd other end
         b.i. blockk.

         S.M.M. N13.

                      1.670
         d/a arch
              2/0.055 = 0.110

             2.040
         arch 0.055
         2/ 2.095 = 4.190
                      5.970

2/5.97   75 x 25 mm Wrot
         Sapele a.d.
         splayed architrave

		Door	0.800
		edge	0.040
		W=	0.840
		Door	2.000
		½/ Top	0.022
			2.022

<table>
<tr><td>2/2/ 0.84
2.02</td><td>Two cts clear
polyurethane
varnish wood
doors   (new wk
Intl</td><td colspan="2"></td></tr>
</table>

S.M.M. V1.2.3.5.
Particulars of materials given in
Preamble in preference to description.

Door dimensions adjusted to include
extra girth of edges – S.M.M. V2.

Glass panel too small to classify
as glazed wood door – no deduction
for glass panel – not exceeding
0.50m²   S.M.M.V2.

Arch		
2/0.025	0.050	
2/0.075	0.150	
2/0.015	0.030	
Frame	0.020	
	0.040	
	0.010	
	0.060	
	0.030	
	0.390	

S.M.M. V5.

= 390 mm

2/ 2.04 0.39	Ditto wood frame & associated mouldings.		
1.67 0.39	(new wk Intl		

Opg.
adj

S.M.M. G26.27.

Deduct overall size of door and frame.

1.67 2.04	Ddt- 75 mm Thk conc block ptn a.b.

S.M.M. T5.

2/ 1.67 2.04	Ddt- Two ct plaster walls a.b.

		Ht 2.040	S.M.M. V4.12.
		Sklg 0.100	
		1.940	Paper not measured behind skirting.
2/1.64	Ddt:		
1.94	Hang pattern		
	wall paper a.b.		
		opg 1.640	S.M.M. F1.3.4.18.21.
		BI 2/0.100 0.200	
		1.840	
1	Precast conc (1:2:4)		Particulars of materials given in
	lintol size		Preamble in preference to description
	1840 x 75 x 150 mm		
	all keyed for		Description quite adequate –
	plaster using wi		bill diagram not required.
	1 No 12 mm dia m.s.		
	bar + bi. blockk		S.M.M. F3.
	in mor (1:2:9).		Classification – other concrete work
			No deduction made on partition.
			Full block course not displaced.
			Block course 225mm high. Lintol
			150mm high. S.M.M. G26.
		opg 1.640	
		arch 2/0.055 0.110	
		o/a 1.780	S.M.M. N13.
2/1.78	Ddt:		
	25 x 100 mm Wrot		
	sapele sklg ab		
	&		
	Ddt:		
	Two cts polyurethane		S.M.M. V4.
	varnish wood sklg		
	n.e. 150 mm gth.		
	(new wk		
	(1HTH		
1.64	Add		S.M.M. N4.
0.08	25 mm Ikk t+g		
	strip floorg a.b.		

114

Plate 13
ALTERNATIVE 'B' GLAZED DOORS

S.M.M. RULES | SECTION A.
| B.10.
| N.1.13.17.19.20.31.32.
| U.3.4.
| V.1.2.3.5.

APPROACH:–

Doors — Glazed doors
Ironmongery
Glass
Frame
Fixing frame
Painting.

Opening
adjustment — Walls and finishings
Head inside and out
Reveals inside and out
Threshold inside and out.

/Doors

Opg 1.670

frame 2/0.040 0.080

2)1.590

£ 0.795

rebate 0.005

W = 0.800

Opg Ht 2.040

frame 0.040

H = 2.000

S.M.M. N1.17.19.

2/ 1 Softwood panelled door size 800 x 2000 x 40 mm thk wi 40 mm thk framg two on solid one side wi rebated meeting stile and 1. No panel rebated and open for glass divided into small panes incl 10 x 20 mm sw. glazing fillets

Particulars of quality given in Preambles in preference to description.

— check size of glass for classification.

115

2/	1	Pair 100 mm pressed steel butts scr to dw.	<u>S.M.M. N32.</u> Usually specified in pairs.
	1	<u>Fix</u> rebated mortice latch w' set lever handle furniture to dw. & <u>Fix</u> 100 mm barrel bolt ditto	<u>S.M.M. N32.</u> If ironmongery is given as a provisional or prime cost sum, the fixing of each unit (or set) is enumerated separately. <u>S.M.M. N32.</u>
	ITEM	Include PRIME COST Sum of £ 10.00 for ironmongery to be supplied by a nominated Supplier & Add for profit	<u>S.M.M. N32. A8. B10.</u> One prime cost sum would be included for the supply of all ironmongery required.

(glass

            door 0.800
stile 0.100
rebate <u>0.010</u>
      2/0.090 = 0.180
glass
bars
  2/0.010   = 0.020 = 0.200
                 <u>3) 0.600</u>
              W = <u>0.200</u>
          Door = 2.000
Top rail 0.100
rebate <u>0.010</u> 0.090
Bot rail 0.150
rebate <u>0.010</u> 0.140
glass
bars 4/0.010 <u>0.040</u> = 0.270
              <u>5) 1.730</u>
          H = <u>0.346</u>

<u>S.M.M. U3.4.</u>

116

$0.200 \times 0.346 = 0.069 \text{m}^2$ —   classification not exceeding 0.10m²

2/
15/ 0.20
     0.35

5mm Thk Deep
Flemish glass &
glazing to wood
w' wood fillets
fxd w' brads in
panes n.e. 0.10 m²
(In No 30).

Following in 1No Door Frame Set:-   <u>S.M.M. N20.</u>

2/ 2.04

100 × 50 mm wrot
Sw. rebated jamb   <u>S.M.M. N17, 20.</u>

1.67

100 × 50 mm ditto
head.

<u>End - Door frame set.</u>

2/ 4

25 × 3 mm Galv m.S.
fxg cramp 150 mm
gth one end
turned up holed
& scr to Sw.
other end b.i.
blockk.   <u>S.M.M. N31.</u>

117

o/a arch   1.670
2/0.055 = 0.110

       2.040
arch   0.055
2/ 2.095 =   4.190
           5.970

2/ 5.94   75 x 25 mm Wrot
       S.W. splayed
       architrave.

S.M.M. V1.2.3.5.

Door dimensions adjusted to include extra girth of edges. S.M.M. V2.

      Door 0.800
      edge 0.040
      W = 0.840

      Door 2.000
½/ 2op 0.020
      Ht 2.020

2/2/ 0.84   K.P.S. ③
2.02   glazed wood door
       in small panes.

Particulars of material given in Preamble in preference to description

Glass size classifies size of pane.

       (new wk
        Intl

      arch
2/0.025   0.050
2/0.075   0.150
2/0.015   0.030
  Frame   0.020
        0.040
        0.010
        0.060
        0.030

S.M.M. V5.

2/ 2.04   Ditto wood   0.390
0.39   frame &
       associated
1.67   mouldings.
0.39

       (new wk
        Intl

       (Opening
       (Adjustment

ALL AS ALTERNATIVE DOORS "A"

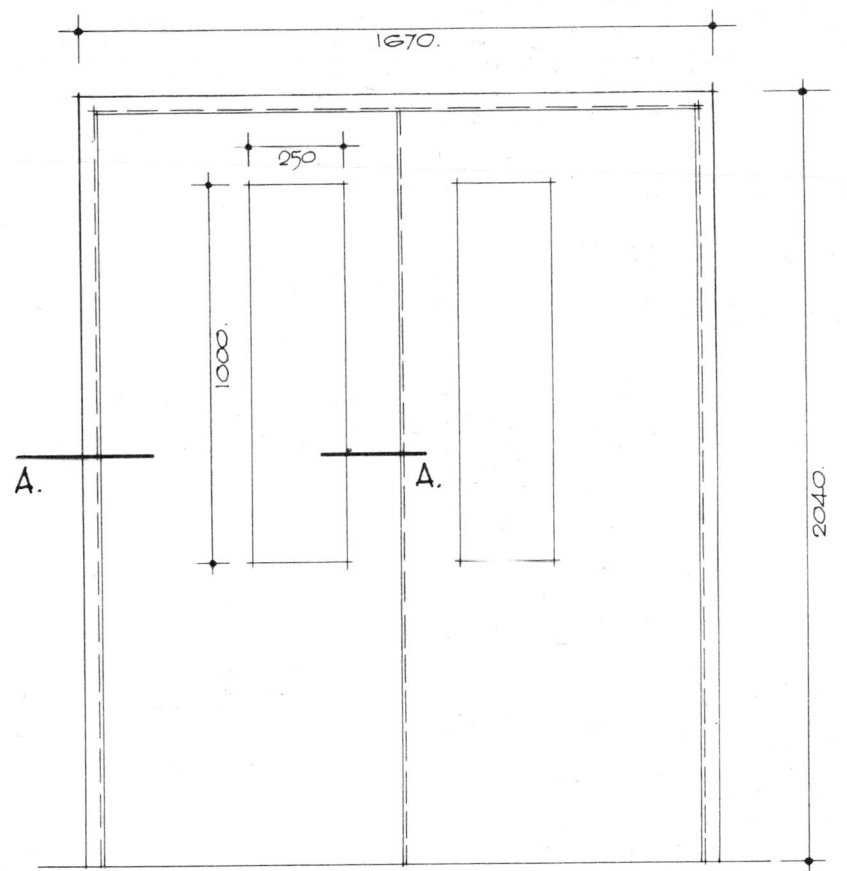

ELEVATION. SCALE 1 : 20.

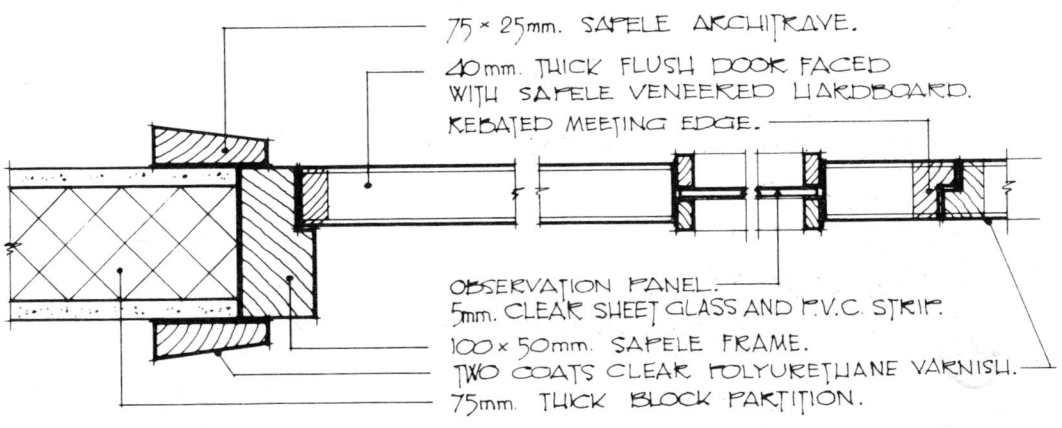

75 × 25mm. SAPELE ARCHITRAVE.

40mm. THICK FLUSH DOOR FACED
WITH SAPELE VENEERED HARDBOARD.
REBATED MEETING EDGE.

OBSERVATION PANEL.
5mm. CLEAR SHEET GLASS AND P.V.C. STRIP.

100 × 50mm. SAPELE FRAME.

TWO COATS CLEAR POLYURETHANE VARNISH.

75mm. THICK BLOCK PARTITION.

PLAN AT A. A. SCALE 1 : 5.

IRONMONGERY :-
PAIR 100mm. PRESSED STEEL BUTTS TO EACH DOOR
REBATED MORTICE LATCH   P.C. £5.00
100mm. BARREL BOLT   P.C. £2.00

Plate 13
'A' FLUSH DOORS

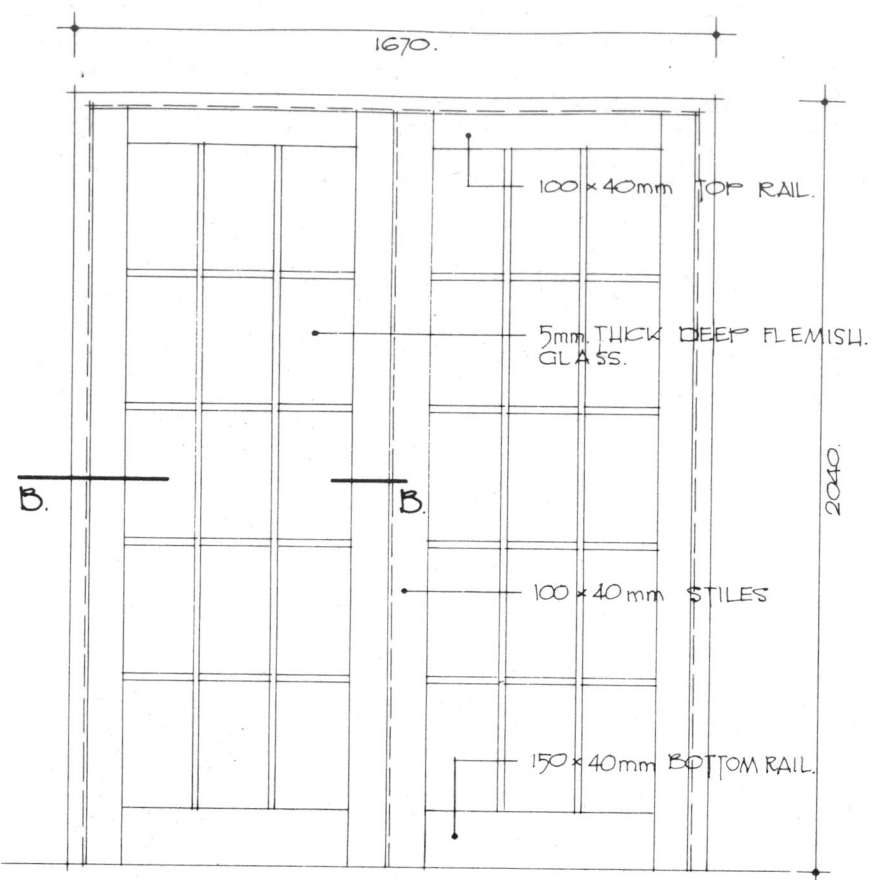

1670.

2040.

100 × 40mm TOP RAIL.

5mm. THICK DEEP FLEMISH. GLASS.

B.

B.

100 × 40 mm STILES

150 × 40mm BOTTOM RAIL.

ELEVATION.   SCALE 1 : 20.

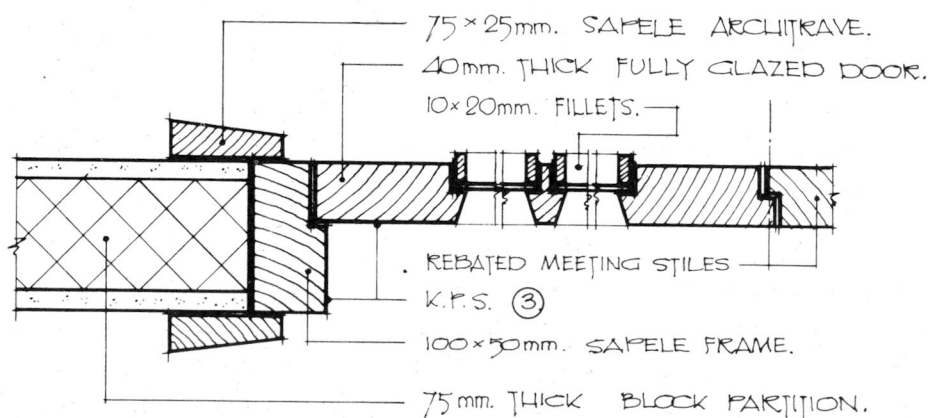

75 × 25mm. SAPELE ARCHITRAVE.

40mm. THICK FULLY GLAZED DOOR.

10 × 20mm. FILLETS.

REBATED MEETING STILES

K.P.S. ③

100 × 50mm. SAPELE FRAME.

75mm. THICK BLOCK PARTITION.

PLAN AT B.B.   SCALE 1 : 5.

IRONMONGERY :-
PAIR 100mm PRESSED STEEL BUTTS TO EACH DOOR
REBATED MORTICE LATCH }   P.C. £10.00
100 mm. BARREL BOLT. }

Chapter IV

Woodwork
Staircases and fittings

Measurement of Staircases and Fittings

STANDARD UNITS. Stock pattern units and standard units shall be enumerated stating the appropriate reference (SMM A6). Stock pattern units are normally identified by a reference number relating to the manufacturer's catalogue and standard units by a reference to the BS or British Woodwork Manufacturer's Association specification.

STAIRCASES. Staircases and short flights of steps shall be enumerated and supported by component details. Landings shall be included with the relevant staircase or flight of steps (SMM N23), where no method of jointing or form of construction is described it shall be deemed to be at the discretion of the Contractor (SMM N1.3). Balustrades shall be enumerated and supported by component details but where forming an integral part of a staircase they should be included with the staircase (SMM N23). Unless included with another item, handrails should be given in metres stating the cross section dimensions (SMM N13).

FLOORS. Adjustments to form openings in floors for staircases are most conveniently made in the floors group while the taker off is still familiar with the floor construction and coverings.

FINISHINGS. The work to staircase areas normally includes the adjustment of any ceiling finishings previously measured over the stairwell and the measurement of additional finishings around the stairwell.

FITTINGS. Items which may be fabricated off-site are defined as composite items (SMM N17) and shall be described in the form they are likely to be delivered to the work. Incorporation into the works is deemed to be included with the items unless otherwise stated. Breaking down into suitable sections for handling purposes, subsequent reassembly and any adjustments necessary to enable the item to be included in the works are also deemed to be included with the items. All metal work, ironmongery and the like included in the description of a composite item is deemed to be fixed.

The supply of composite items shall be enumerated and supported by component details, where no method of jointing or form of construction is described it shall be deemed to be at the discretion of the Contractor (SMM N1.3). Alternatively, fittings shall be the subject of a provisional sum (SMM N26). Provisional sums must be included if component details are not available at the time of preparation of the bills of quantities. The fixing of such fittings shall be enumerated and given separately (SMM N27). Where fittings are the subject of a provisional sum such sum shall be deemed to include fixing (SMM N27).

The very small number of fittings which might be fabricated on-site are excluded from the definition of composite items (SMM N17). Such fittings as isolated shelves, work-tops, seats, pelmets, pipe-casings, hat and coat rails and the like should be measured in detail in accordance with SMM N13.

COMPONENT DETAILS AND BILL DIAGRAMS. It is important to distinguish between component details and bill diagrams (SMM A5). Component details must show all the information necessary for the manufacture and assembly of the component. Bill diagrams are usually simple diagrams which may be provided to aid the descriptions contained in the bill of quantities. Component details would normally be provided by the Architect and bill diagrams by the Quantity Surveyor.

PAINTING. All painting and decorating work on staircases and other fittings is best measured together after each composite item since many of the members are not painted in isolation but in conjunction with each other (SMM V4). Work to staircase areas is classified separately to allow the estimator to price working off staircase flights and/or in restricted areas (Practice Manual V.4.1c).

Plate 14
STRAIGHT FLIGHT STAIRCASE

<u>S.M.M. RULES</u>	SECTION A. N.1.5.7.13.17.23. Q.5. V.1.2.3.4.	
<u>APPROACH:-</u>	Staircase Handrails Apron lining - none Spandril filling - none Painting ―――――――	
1	Wrot s.w. straight flight staircase 900 mm wide 2660 mm total rise as <u>Component Detail No 14.</u>	<u>S.M.M. N23.</u> Plate No.14 to be included as Component Detail.
0.87	90 x 25 mm Wrot s.w. nosing tongued to edge of floorg incl groove.	<u>S.M.M. N5.7.</u> Nosing No.14 - Measured with staircase but billed with flooring. Ensure that nosing is not also measured with the floor.
2	Ends ditto	<u>S.M.M. N1.</u>
4.13	50 mm dia wrot hardwood mopstick handrail scr to bkts.	<u>S.M.M. N13.</u> Not integral part of staircase.
2	Radod end	<u>S.M.M. N1.</u>

mild steel
handrail bracket
150 mm gth
75 mm projection
plugged & scr
to bkk.

tread	0.225
nosg	0.015
thk	0.032
proj	0.015
	0.287
rise	0.190
tread	0.032
	0.158
	0.225
	0.070
2)	0.295
Avg	0.148
string	0.032
plaster	0.012 0.020
	0.168

S.M.M. V1.2.3.4.

Painting measured 300mm wide margins
on treads and risers and to exposed
faces of strings.

2/14/ 0.30	
0.29	
2/14/ 0.30	
0.16	
2/ 4.25	
0.17	

K.P.S. ③
staircase
areas.

(new wk
1/4t2

— Treads including top nosing

— Risers

— Strings

4.13

Ditto n.e.
150 mm gth
(new wk
1/4t2.

— Handrail

Appropriate adjustment to floors
and finishings to follow here.

124

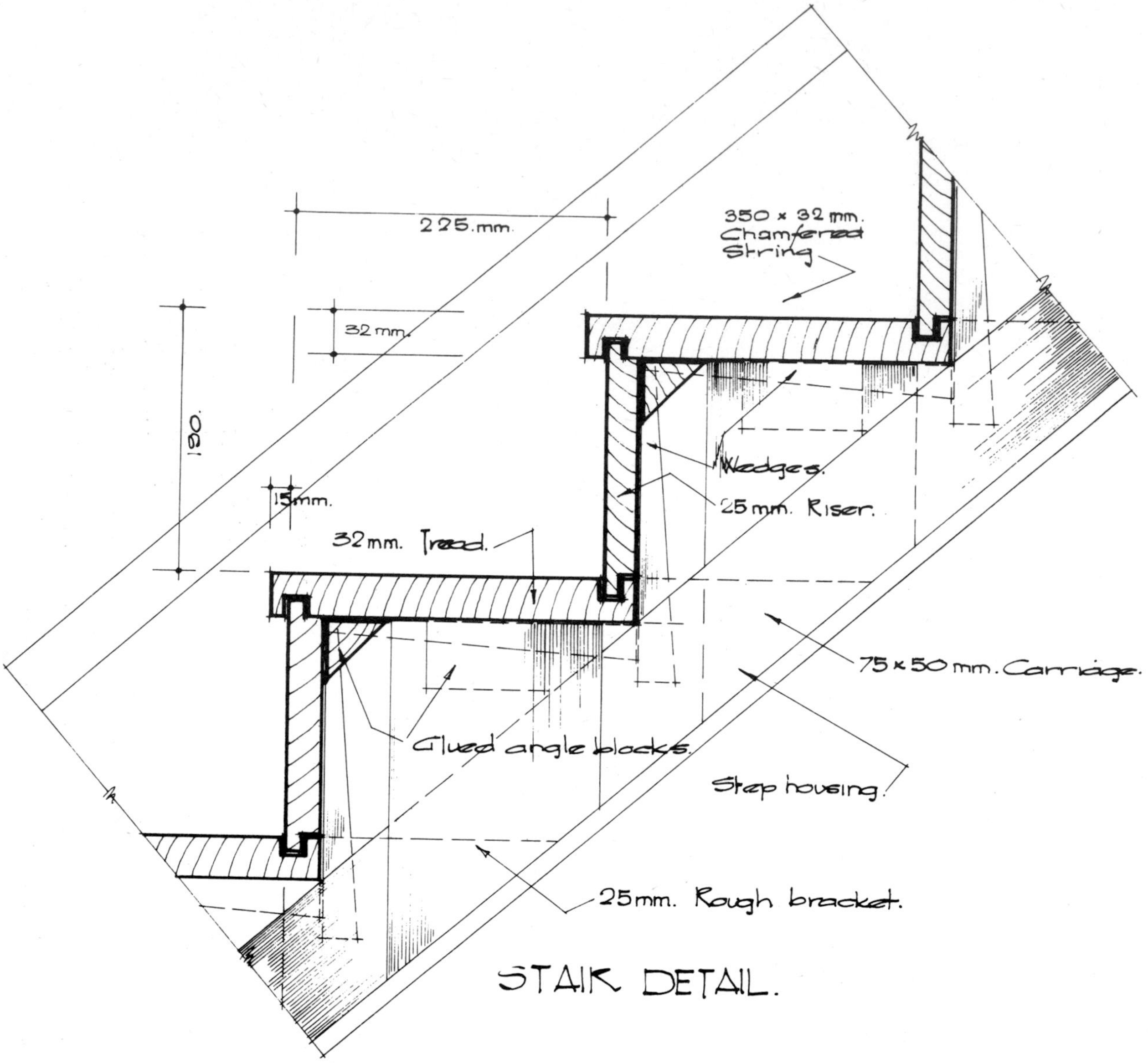

225 mm.

32 mm.

350 × 32 mm. Chamfered String

180

15mm.

32mm. Tread.

Wedges.

25 mm. Riser.

75 × 50 mm. Carriage.

Glued angle blocks.

Step housing.

25mm. Rough bracket.

STAIR DETAIL.

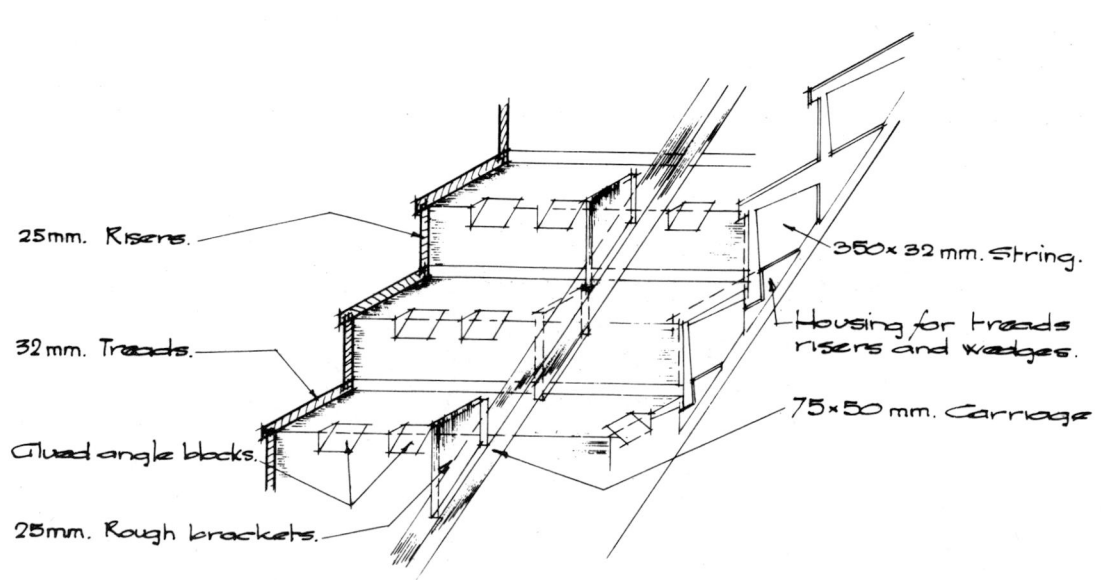

25mm. Risers.

32 mm. Treads.

Glued angle blocks.

25mm. Rough brackets.

350 × 32 mm. String.

Housing for treads risers and wedges.

75 × 50 mm. Carriage

Plate 14
STRAIGHT FLIGHT STAIRCASE

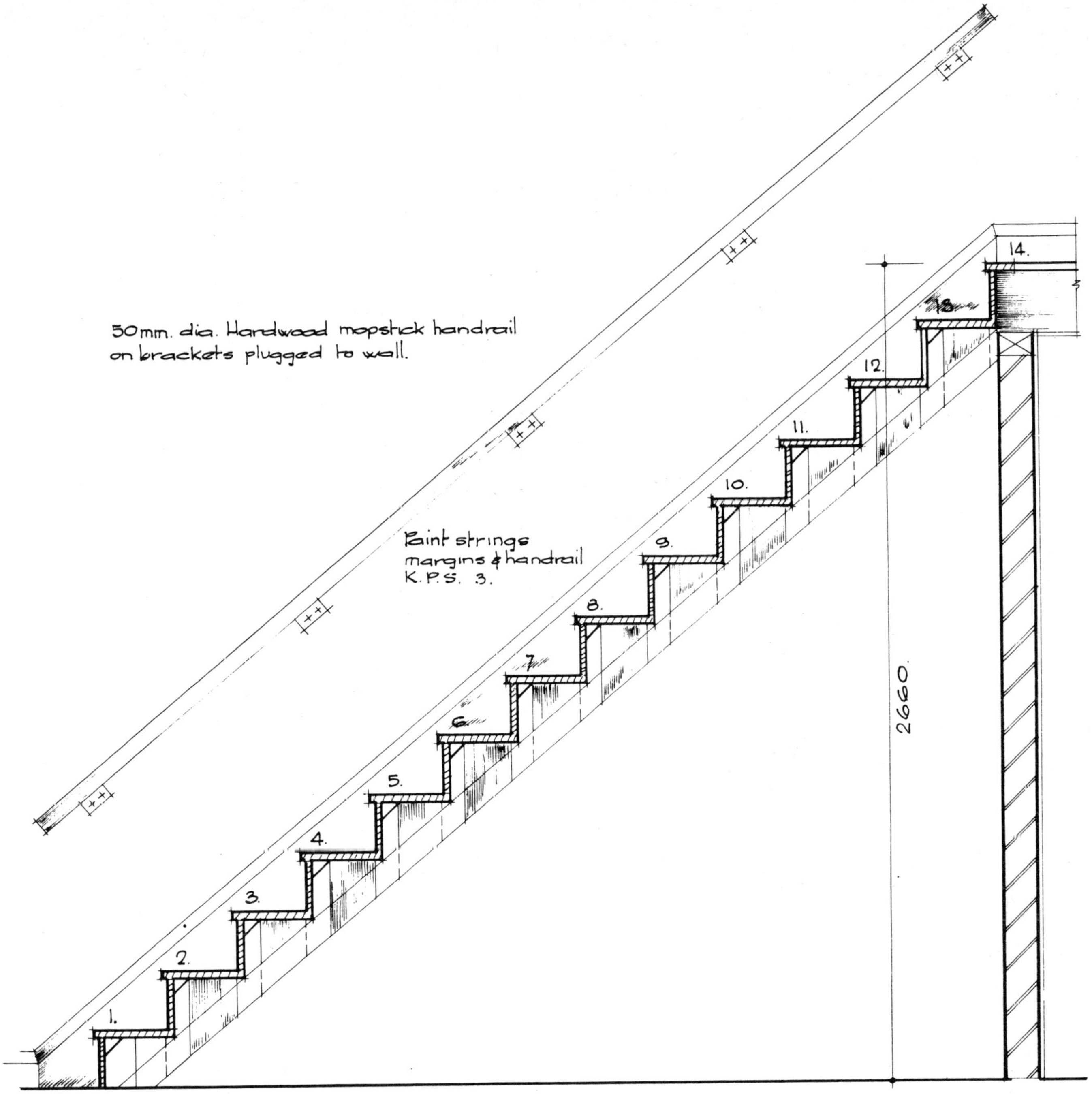

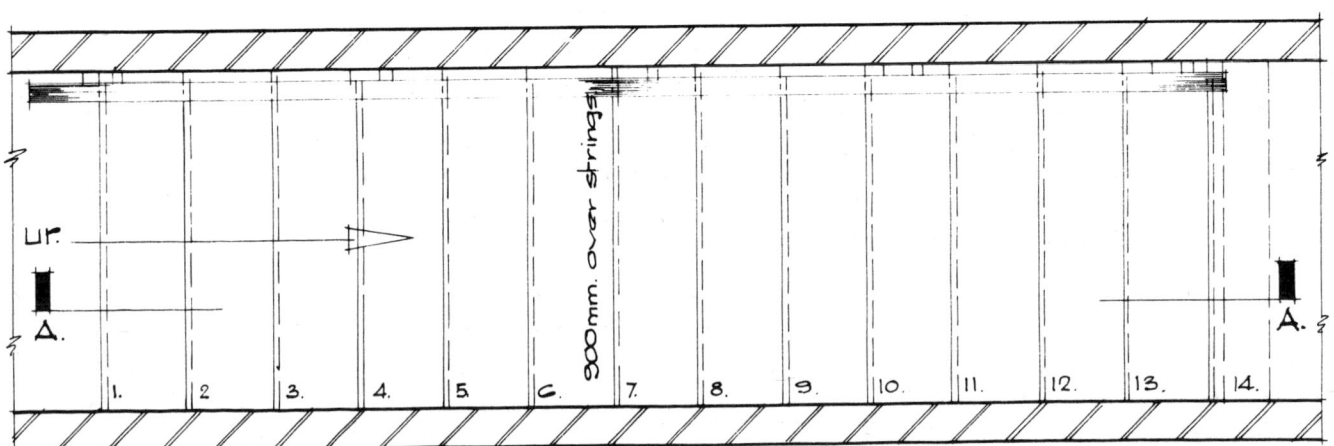

50mm. dia. Hardwood mopstick handrail
on brackets plugged to wall.
Paint strings
margins & handrail
K.P.S. 3.
1.
2.
3.
4.
5.
6.
7.
8.
9.
10.
11.
12.
14.
2660

SECTION A.A.

UP.
A.
A.
900mm. over strings.
1.
2
3.
4.
5.
6.
7.
8.
9.
10.
11.
12.
13.
14.

PLAN.

Scale 1 : 20.

Plate 15
DOG LEG STAIRCASE

S.M.M. RULES	SECTION A.
	     N.1.5.7.11.13.17.23.24.
	     V.1.2.3.4.		

APPROACH:-

Staircase
Landings   ) Integral part of staircase.
Balustrade   )
Apron linings
Spandril filling – none
Painting

---

1   Wrot s.w. dog leg staircase of two straight flights wi half space landing & panelled balustrade as

COMPONENT DETAIL

No 15.

          S.M.M. N23.24.

          – Plate No.15 to be included as Component Detail.

1·06
1·80
0·90
     25 × 100 mm s.w. sklg a.b.

          S.M.M. N13.
          Measured with staircase for convenience but billed with other skirtings.

2   End ditto

        &

     mitre ditto

          S.M.M. N1.

	0.800	
	Loused ends	
	2/0.015 0.030	
	⎯⎯⎯⎯	
	0.830	
0.83	75 × 30 mm	Floor nosing No.15.
	wrot sw. nosg	Measure with staircase but
	tongued to edge of	billed with flooring.
	floorg incl groove	Ensure that nosing is not also measured with the floor.
2	Ends ditto	S.M.M. N1.
		S.M.M. N5.%.
0.80	100 × 30 mm wrot	First floor.
	sw. nosg tongued	Ensure that nosing is not also
	to edge of floorg	measured with the floor.
	incl groove.	
2	Ends ditto	S.M.M. N1.
		S.M.M. N1.13.
0.80	20 mm thk wrot	
	sw. apron lining	
	225 mm wide	
	in one width	
	&	
	50 × 100 mm sawn	S.M.M. N11.
	sw. batten	
2	Ends apron lining	S.M.M. N1.

				tread	0.250		S.M.M. V1.2.3.4.
					0.025		Painting measured 300mm wide to
					0.030		margins on treads and risers.
					0.025		
					0.330		All painting measured in square
							metres. ie. not isolated surfaces.

```
2/2/                          K.P.S. ③ staircase
 7/ 0.30                      areas.   ( new wk        –   Treads
    0.33                               ( in 12

2/8/
 7/ 0.30                                               –   Risers
    0.15
                                       sktg   1.800
                                   2/0.035   0.070
                                             1.730
                                             1.125
                                             1.075
                                          2) 2.200
                                             1.100
                                       sktg  0.035
                                             1.065
    1.73                                               –   Landing
    1.07

    0.20                                               –   Quadrant end
    0.15
                                             0.055
                                             0.215
                                          2) 0.270
                                       Avg   0.135
                                       2op   0.035
                                             0.170      –   Wall strings
    2.57
    0.17
    2.25
    0.17
    1.06                                               –   Landing skirting
    0.13
    1.80
    0.13
    0.90
    0.13
```

a.b. Avg 0·135
top 0·015
0·150

2/ 2·10
0·15

K.P.S. ③ staircase areas. (new cube 1 N/2

      — Outer string (inside face).

0·250
0·015
0·265

2/ 2·10
0·27

     — Outer string (outside face).

1·30
0·40

     — Newels

2·65
0·40

1·40
0·40

1·40
0·20

1·20
0·18

     — Handrail

1·84
0·18

0·90
0·18

2/ 1·40
0·57

     — Balustrades

2/ 1·74
0·63

2/ 0·80
0·82

2/ 0·80
0·65

     — Plate

apron 0·020
0·225

nosg
2/ 0·020 0·040
0·030

0·315

     — Apron lining and edge of floor.

0·80
0·32

Appropriate adjustment to floors and finishings to follow here.

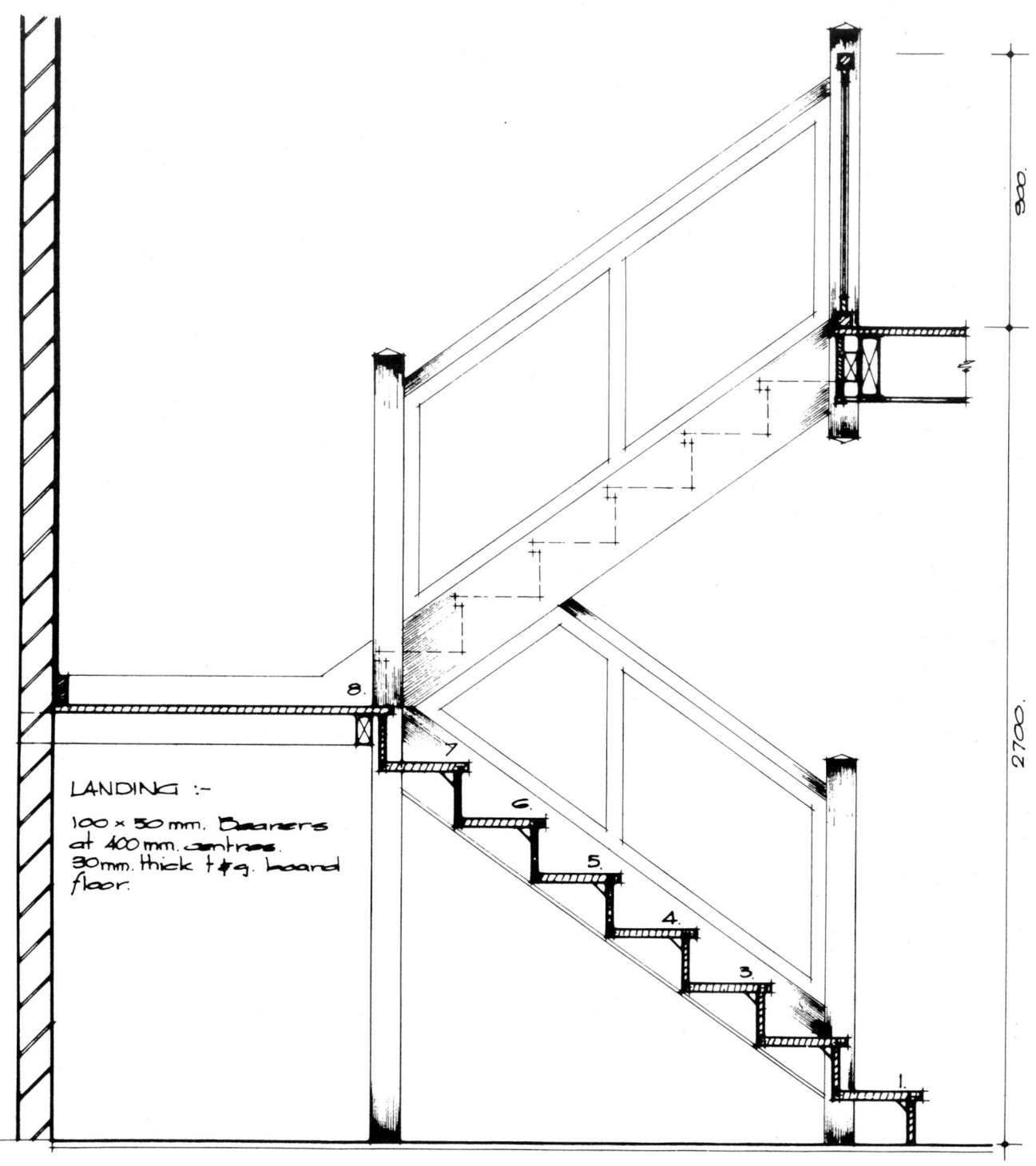

LANDING :-

100 x 50 mm. Bearers
at 400 mm. centres.
30 mm. thick t & g. board
floor.

SECTION A. A.

Scale 1 : 20.

Plate 15
DOG LEG STAIRCASE

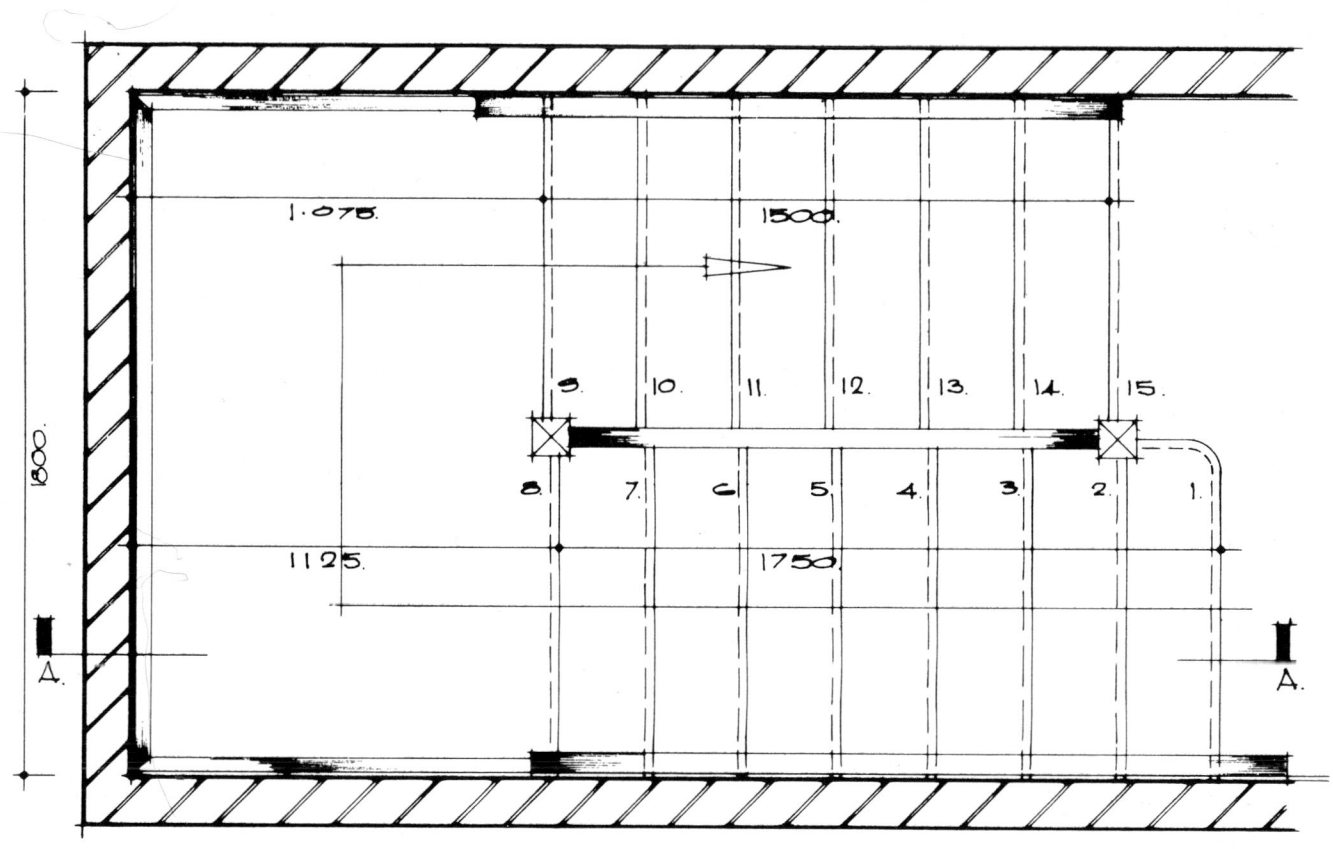

PLAN.

Scale. 1 : 20.

SPECIFICATION.

50 × 50 mm. Handrail. 20 mm. thick panelled balustrade
with 12 mm. thick blockboard panels.
250 × 50 mm. Strings. 30 mm. thick treads and nosings.
25 mm. thick risers.
100 × 100 mm. newels with balustrade moulds planted on.
K. P. S.   ③.

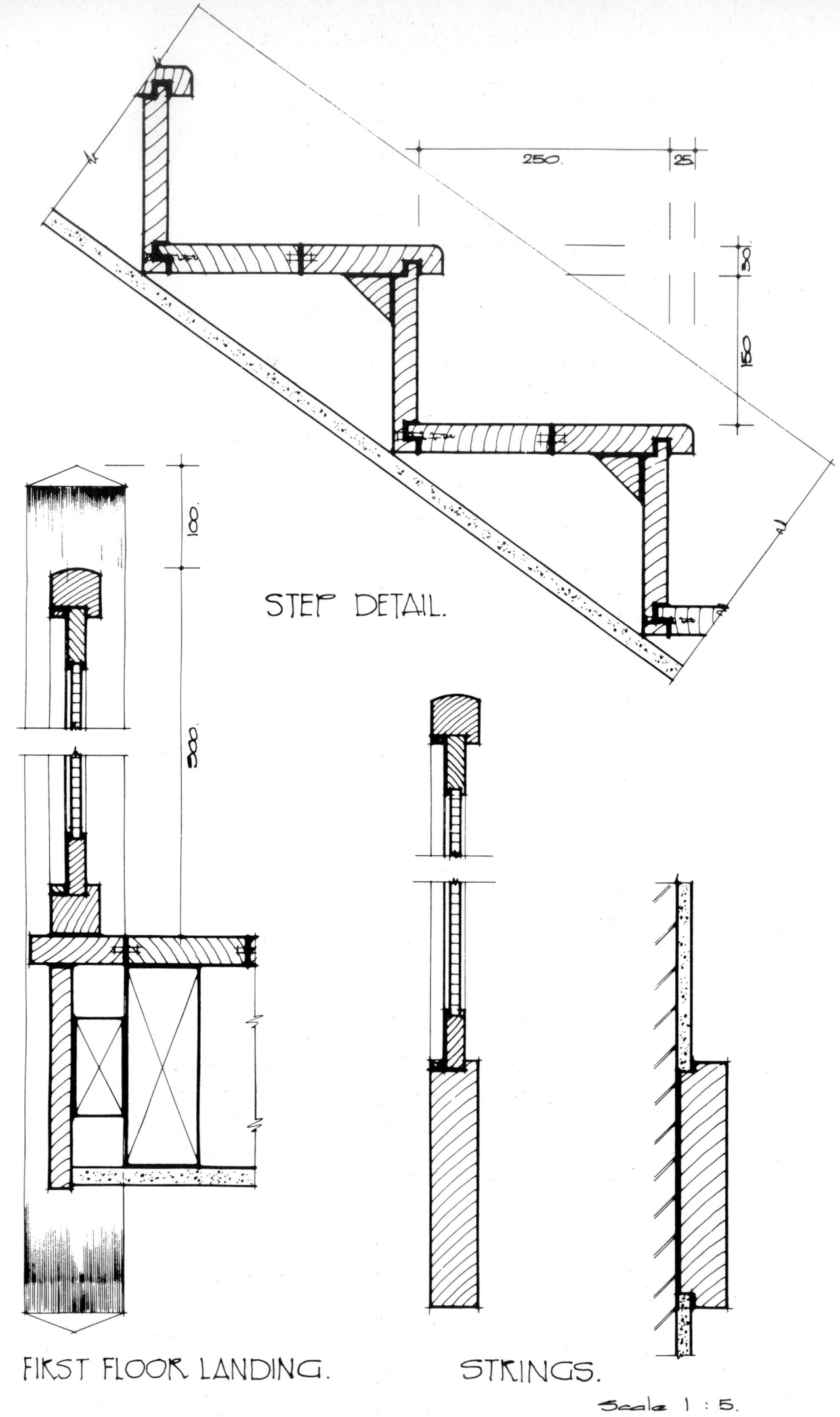

250.   25

50

150

STEP DETAIL.

FIRST FLOOR LANDING.   STRINGS.

Scale 1 : 5.

Plate 16
OPEN NEWEL STAIRCASE

S.M.M. RULES   SECTION A.
                   N.1.17.23.24.
                   V.1.2.3.4.

APPROACH:–
     Staircase
     Landings   ) Integral part of staircase.
     Balustrade   )
     Apron linings
     Spandril filling – none
     Painting.

		S.M.M. N23.24.
1	Wrot S.W. open newel staircase of two straight flights wi' set winders and balustrade as COMPONENT DETAIL No. 16.	Plate No.16 to be included as Component Detail.
	string 0.030 0.900 ½ newel 0.050 0.080 string – newel = 0.820 housed ends 2/0.015 0.030 0.850	S.M.M. N5.7.
0.85	75 x 30 mm wrot S.W. nosg tongued to edge of floorg incl groove.	Nosing No.14 Measured with staircase but billed with flooring. Ensure that nosings are not also measured with the floor.
2	Ends ditto	S.M.M. N1.

0·91
─────
2·10

40 × 40 mm s/w.
2ce rebated
cover fillet

S.M.M. N13.

&

75 × 30 mm Wrot
s/w nosg ab

S.M.M. N5.%.

2/ 2

Ends nosg

S.M.M. N1.

1

notchg board
floorg 250 mm
gth

S.M.M. N5.

&

Ditto 360 mm gth

&

Ditto 130 mm gth

0·91
─────
2·10

25 mm Ък wrot
s/w. apron lining
200 mm wide

S.M.M. N13.

2/ 2

Ends ditto

S.M.M. N1.

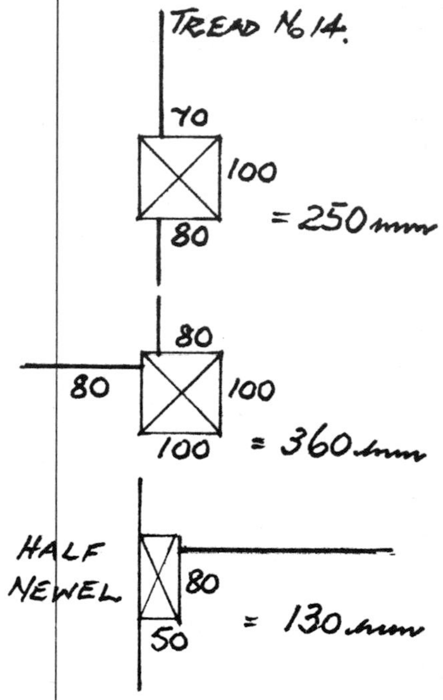

		Tread 0·210
		Nosg 0·020
		Ink 0·030
		Nosg 0·020
		0·280

S.M.M. V1.2.3.4.
Painting measured 300mm wide to margins on treads and risers.

All painting measured in square metres. ie. not isolated surfaces.

Timesing	Dim.	
2/11	0·30	K.P.S. ③ staircase
	0·28	areas. (new wk INTL
2/14	0·30	— Treads
	0·16	
	0·94	— Risers
	0·93	
	0·20	— winders
	0·16	
	1·50	— Quadrant end
	0·14	
	1·95	— Wall string
	0·14	
2/1	1·10	— Wall string
	0·14	
	1·03	— Wall string
	0·12	
	1·89	— Outer string (inside face).
	0·12	
	0·74	— Outer string (inside face).
	0·50	
	1·63	— Outer string (outside face).
	0·50	
	1·34	— Outer string (outside face).
	0·40	
	2·55	— Newels
	0·40	
2/	1·41	— Newels
	0·40	

Calculations (centre column):

0·200
0·045
2)0·245
Avg 0·123

2ops 0·030
Plaster 0·013 : 0·017
0·140

— Newels

50
50 / 50
10   10
225
10
40 / 10
40
= 495 mm gth

1.41	Ditto n.e. 150 mm gth. (new wk INTZ	–	S.M.M. V1.2.3.4. Half newel. Isolated surface not exceeding 150mm girth.
0.81 1.69 1.01 2.20	Ditto n.e. 300 mm gth. (new wk INTZ	–	Handrails
14/0.73 28/0.85	Ditto n.e. 150 mm gth (new wk INTZ	–	Balusters
0.91 0.50	Ditto staircase areas ab.	–	Plate
2.10 0.50	(new wk INTZ.	–	Apron lining etc.

50
40 ▨ 40
5
30
30
200
10
40 ▨ 15
40
= 500 mm gth

Appropriate adjustment to floors and finishings to follow here.

138

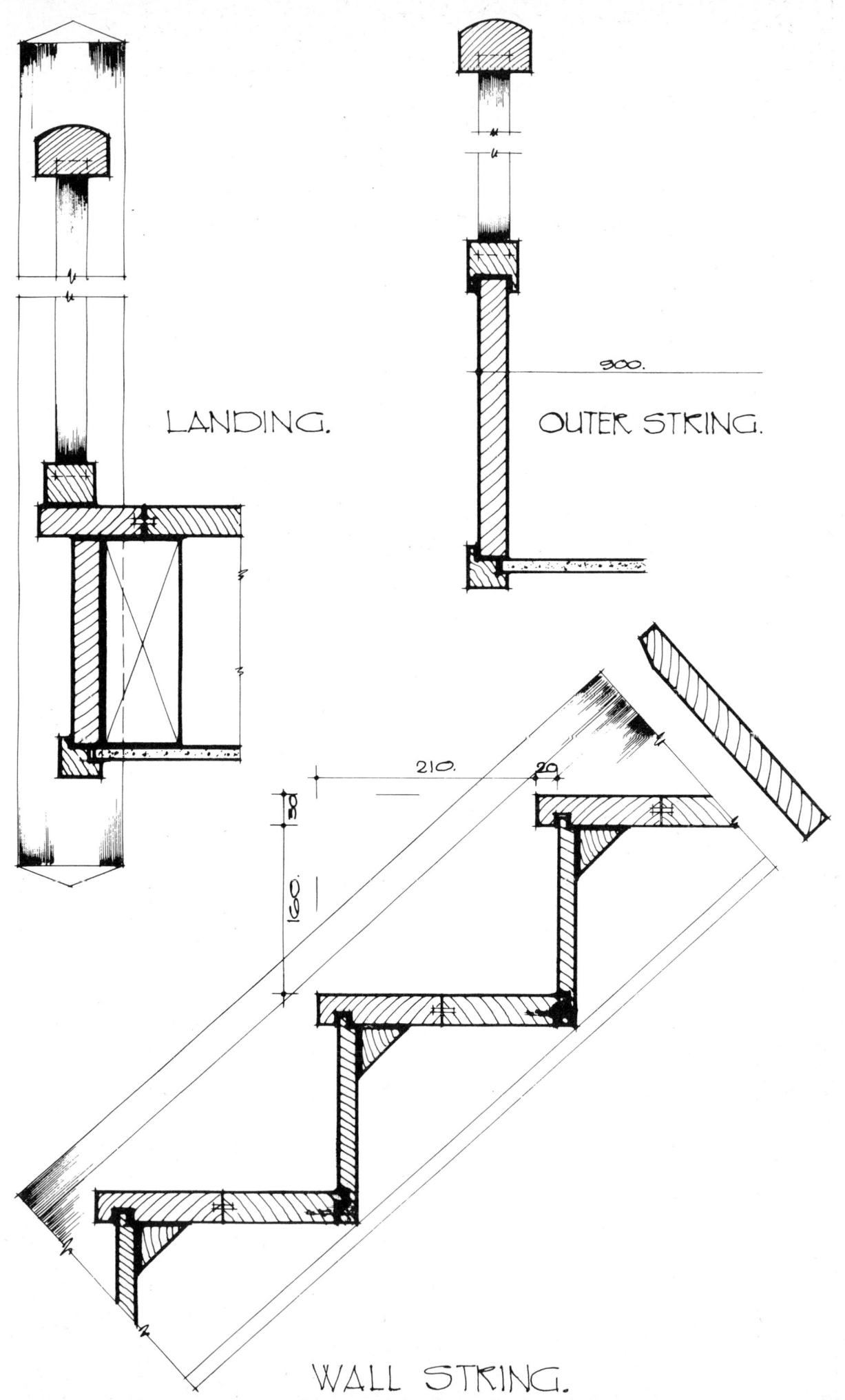

LANDING.

OUTER STRING.

900.

210.

20

50

180

WALL STRING.

Scale. 1 : 5.

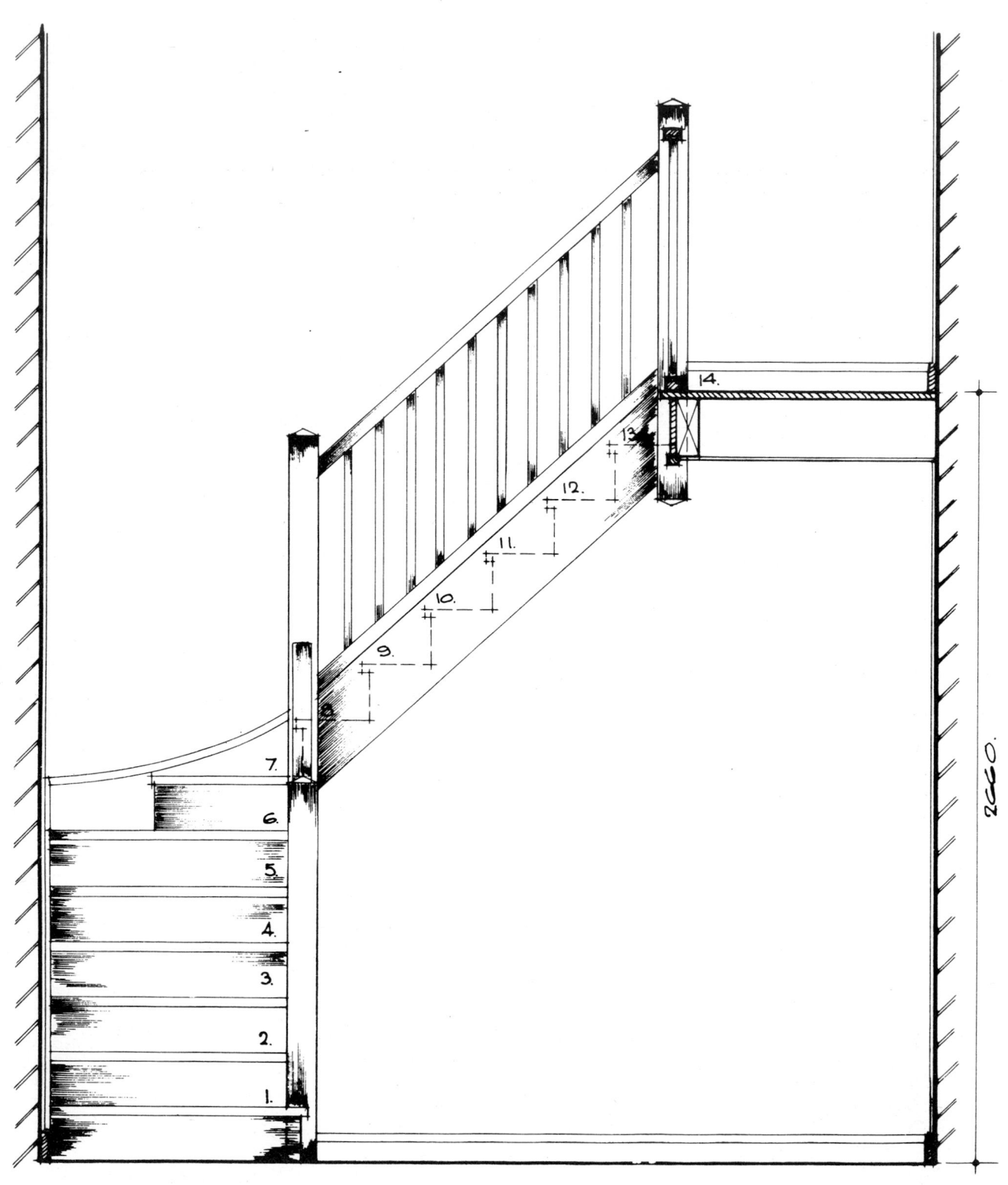

SECTION A. A.

Plate 16
OPEN NEWEL STAIRCASE

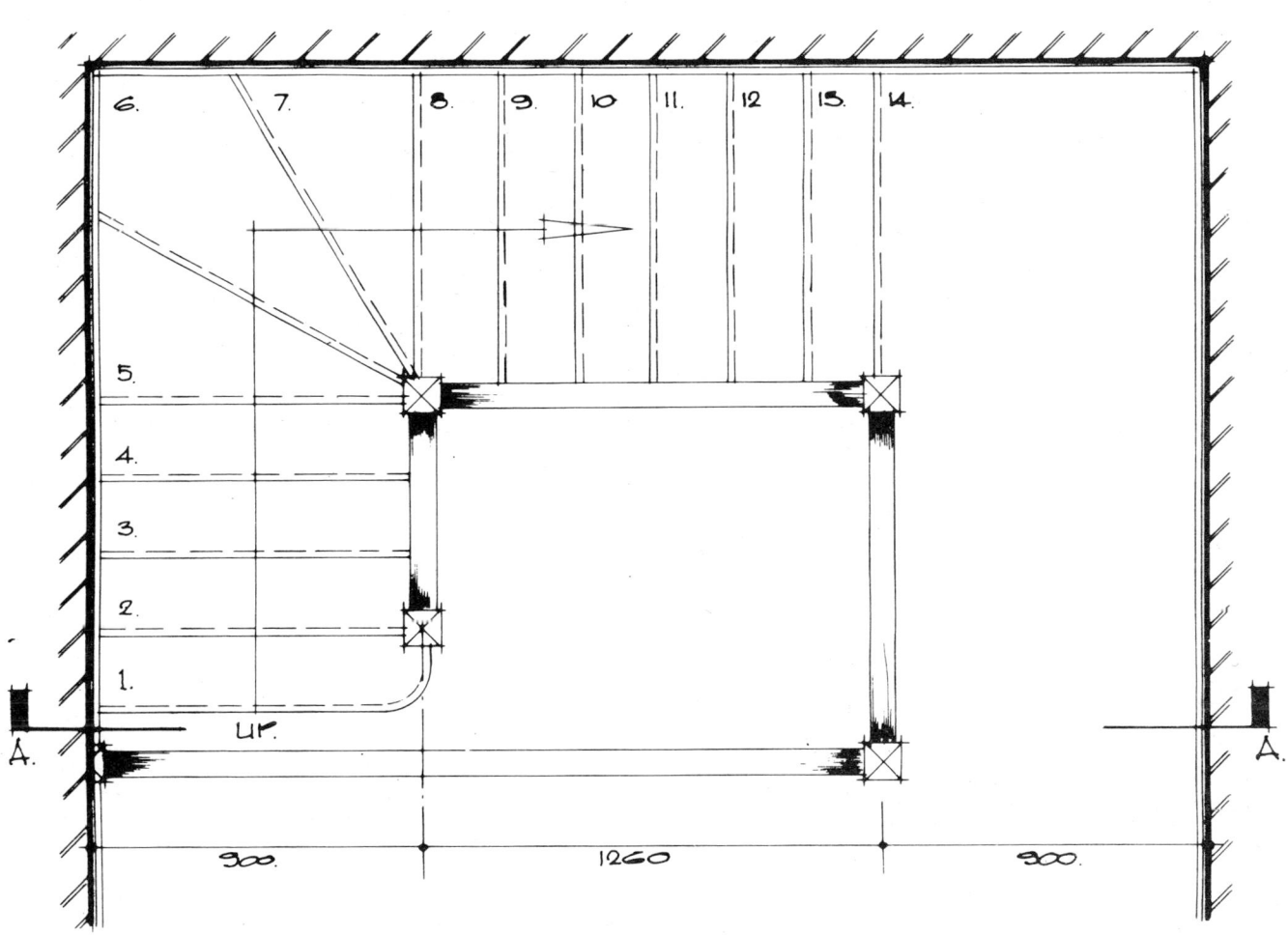

PLAN.   Scale 1 : 20.

SPECIFICATION.

30mm. Thick treads and nosings. 20mm. thick risers.
250 × 30mm. strings. 100 × 100 mm. newel posts.
30 × 30 mm. Balusters. 70 × 50mm handrails.
50 × 50mm. string capping. 40 × 40mm. cover mould to soffit.
K.P.S. ③

Plate 17
STOCK PATTERN UNITS

S.M.M. RULES	SECTION A.
	N.1.17.26.27.

APPROACH:- The requirements of the S.M.M. are deemed
to have been complied with if the item concerned
is a product details of which have been
published and to which reference has been made
in the description (S.M.M. A.6.)

SUPPLY THE FOLLOWING		S.M.M. A6. N17.26.
ELIZABETH ANN "CALYPSO"		Units supplied with decorative
RANGE KITCHEN FITTINGS		finish and all necessary ironmongery.
MANUFACTURED BY		
ELIZABETH ANN WOODCRAFT LTD.,		
RHYL. NORTH WALES :-		

1	Stainless steel sink unit Ref. 21/2.	
	&	
	Corner Unit Ref. 21/5 RH.	- Appropriate reference from manufacturer's catalogue
	&	
	Double Base Unit Ref. 21/3 S.	
	&	
	Broom Cupd. Ref 21/143.	

	FIX ELIZABETH ANN KITCHEN FITTINGS, PLUG & SCREW TO BRICKWORK :-		S.M.M. N27.
1	Sink Unit Ref 21/2 & Corner Unit Ref 21/5 RH & Double Base Unit Ref 21/3 S. & Broom Cupd Ref 21/143.		
END OF "FIX KITCHEN UNITS".			

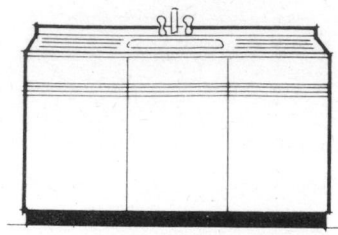

STAINLESS STEEL SINK UNIT REF. 21/2.
1600mm. × 533mm. × 914mm. HIGH.

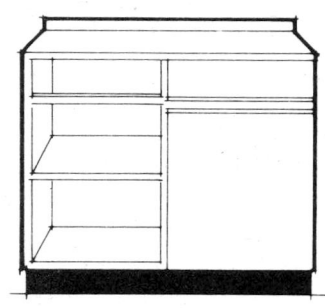

CORNER UNIT.   REF. 21/5 R.H.
1067mm × 533mm × 914mm HIGH.

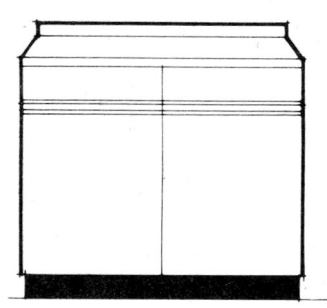

DOUBLE UNIT   REF. 21/3S.
914mm × 533mm × 914mm HIGH.

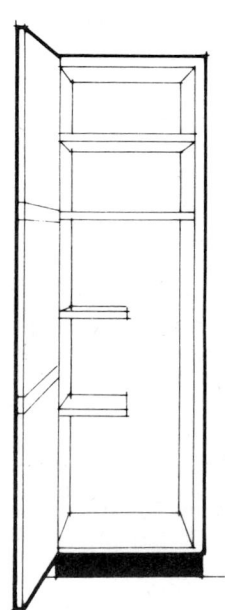

BROOM CUPBOARD.   REF. 21/143.
533mm × 533mm × 2007mm. HIGH.

Plate 17
STOCK PATTERN UNITS

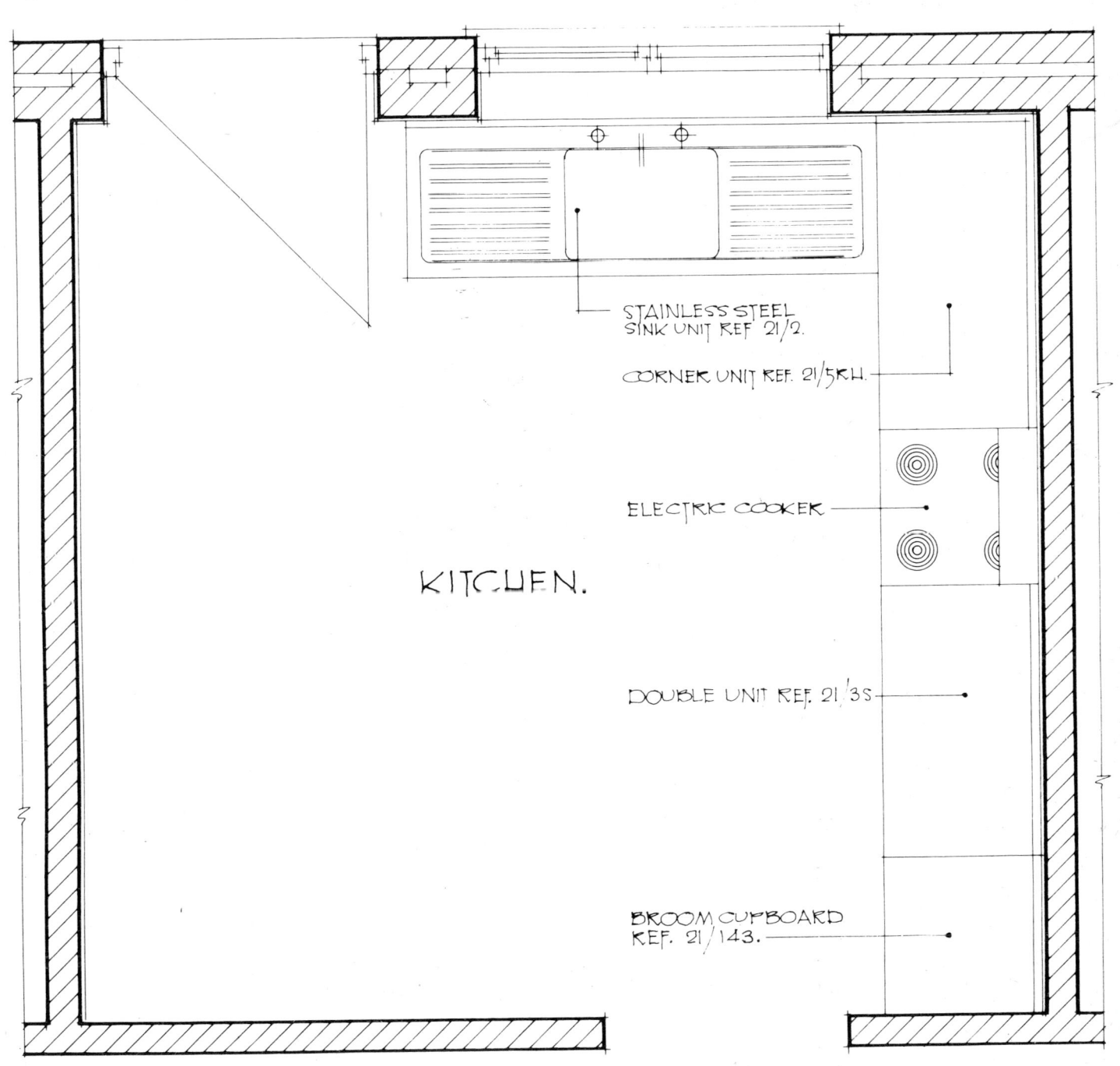

STAINLESS STEEL
SINK UNIT REF. 21/2.

CORNER UNIT REF. 21/5RH.

ELECTRIC COOKER

KITCHEN.

DOUBLE UNIT REF. 21/3S

BROOM CUPBOARD
REF. 21/143.

PLAN.   SCALE 1 : 20.

ELIZABETH ANN "CALYPSO" RANGE FITTINGS.
ELIZABETH ANN WOODCRAFT LTD. RHYL NORTH WALES.

Plate 18
SIMPLE COUNTER

S.M.M. RULES

SECTION A.
N.1.17.26.27.

APPROACH:-

Supply of fittings shall be enumerated and
supported by component details. Alternatively
they shall be the subject of a provisional
sum. (S.M.M. N.26).

1	Supply counter fitting size 2000 x 1000 x 750 mm overall as COMPONENT DETAIL No 18.	S.M.M. N26.
		Plate No. 18 to be included as Component Detail.
	&	
	Fix ditto, placed into position.	S.M.M. N27.

ALTERNATIVE MEASUREMENT IF
COMPONENT DETAILS NOT AVAILABLE :-

ITEM	Include the provisional sum of £350.00 for the supply of Counter Fitting.	S.M.M. N26.27.
		Fixing deemed to be included.

1000.

CENTRE LINE OF COUNTER —

HALF FRONT ELEVATION.

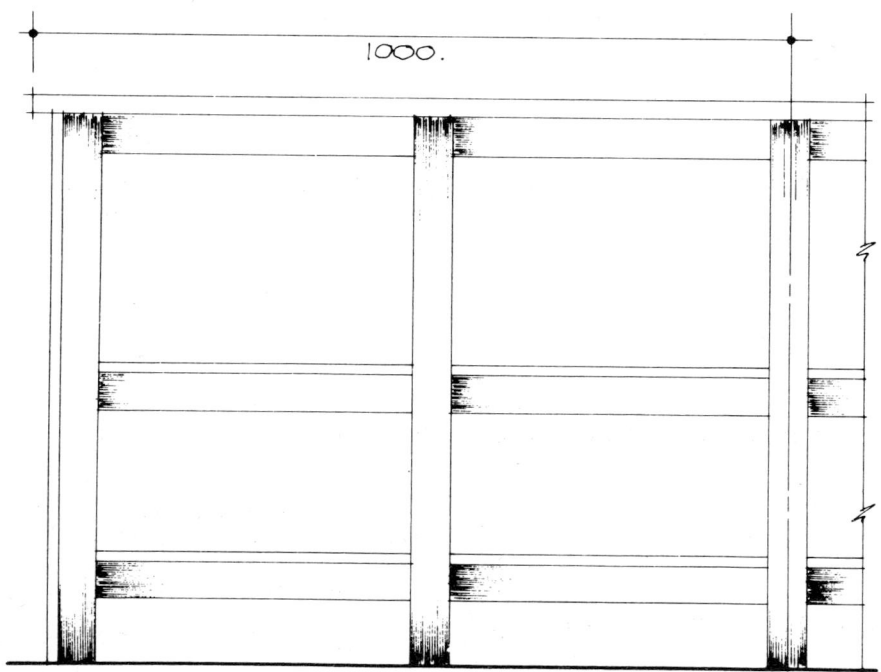

1000.

HALF REAR ELEVATION.

Plate 18
SIMPLE COUNTER

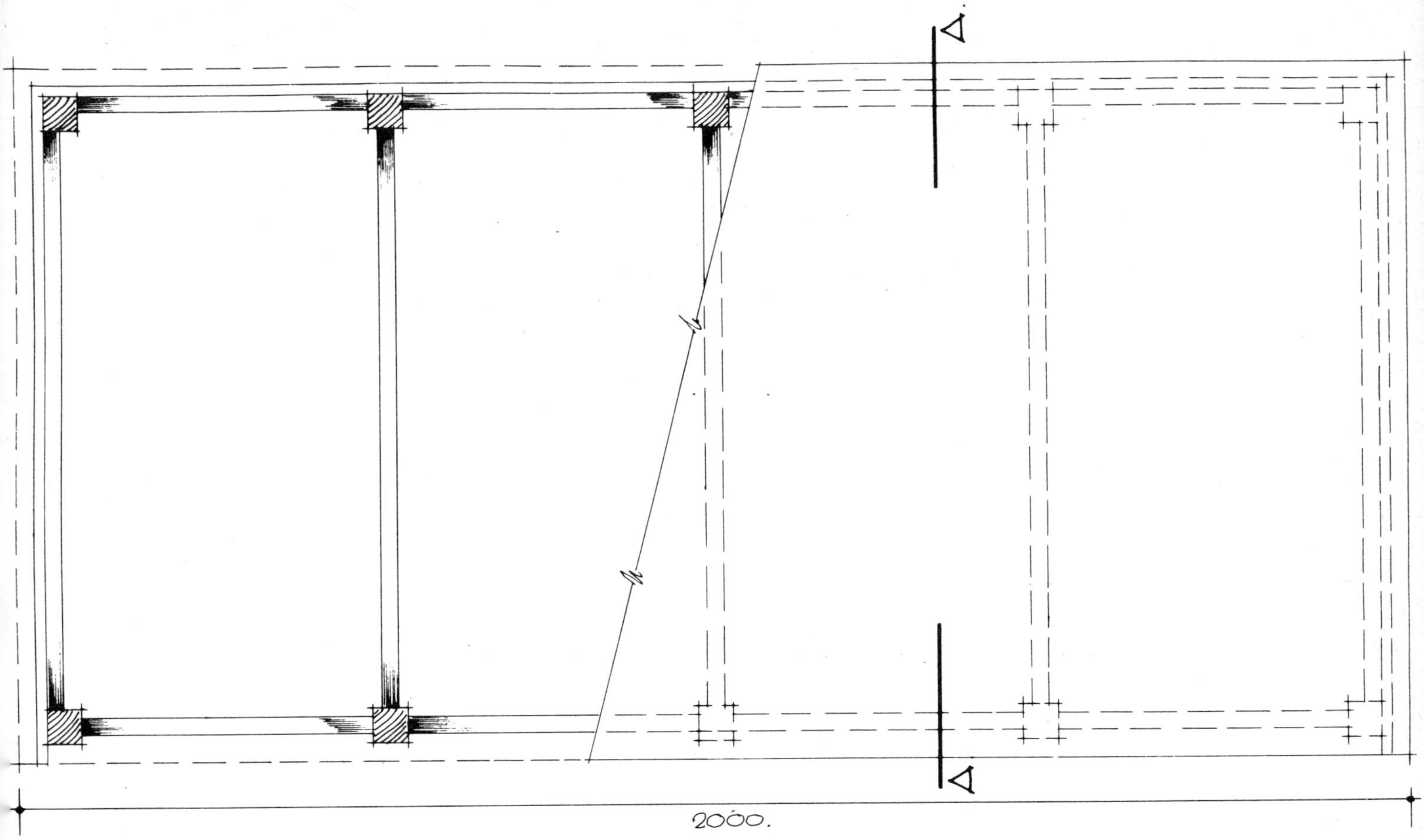

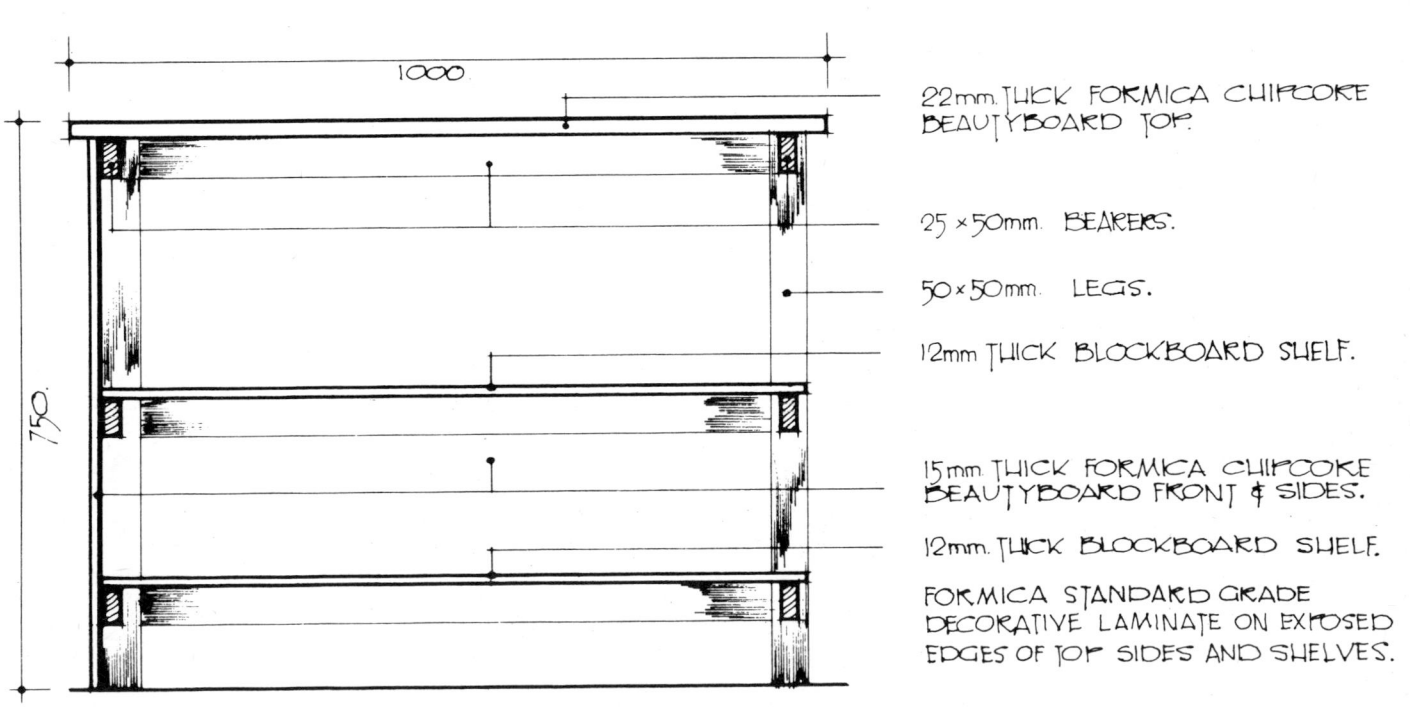

2000.

PLAN.

1000.

750.

22mm. THICK FORMICA CHIPCORE
BEAUTYBOARD TOP.

25 × 50mm. BEARERS.

50 × 50mm. LEGS.

12mm THICK BLOCKBOARD SHELF.

15mm THICK FORMICA CHIPCORE
BEAUTYBOARD FRONT & SIDES.

12mm. THICK BLOCKBOARD SHELF.

FORMICA STANDARD GRADE
DECORATIVE LAMINATE ON EXPOSED
EDGES OF TOP SIDES AND SHELVES.

SECTION A.A.   SCALE 1 : 10.

Plate 19
LOCKER UNIT

S.M.M. RULES		SECTION A.	
		N.1.17.26.27.	
APPROACH:-		Supply of fittings shall be enumerated and supported by component details. Alternatively, they shall be the subject of a provisional sum. (S.M.M. N.26).	

1	Supply locker unit size 2854 × 375 × 1666 mm Overall as COMPONENT DETAIL No 19.	S.M.M. N26.
		Plate no.19 to be included as Component Detail.
	& Fix ditto, placed in position.	S.M.M. N27.

S.M.M. V1.2.3.4.
Particulars of materials given in Preamble in preference to description.

2·85	K.P.S. ③ g.w.s. n.e. 300 mm gth (new wk Intz		
	edge	0·375 0·022	
		0·397	— Top
2·85 0·40	Ditto g.w.s.		— Sides
2/1·67 0·40	(new wk Intz		

150

5/ 1.55			] Ditto n.e. 150 run gth		–	Edge – vertical divisions		
6/ 0.46			(new wk 1 nr72		–	Edge – horizontal divisions		

$$\begin{array}{r} 0.450 \\ edge \quad 0.019 \\ \hline 0.469 \\ 0.450 \\ 0.019 \\ \hline 0.469 \end{array}$$

12/ 2/ 0.44			Ditto g.w.s.		–	Doors		
0.44			(new wk 1 nr72.					

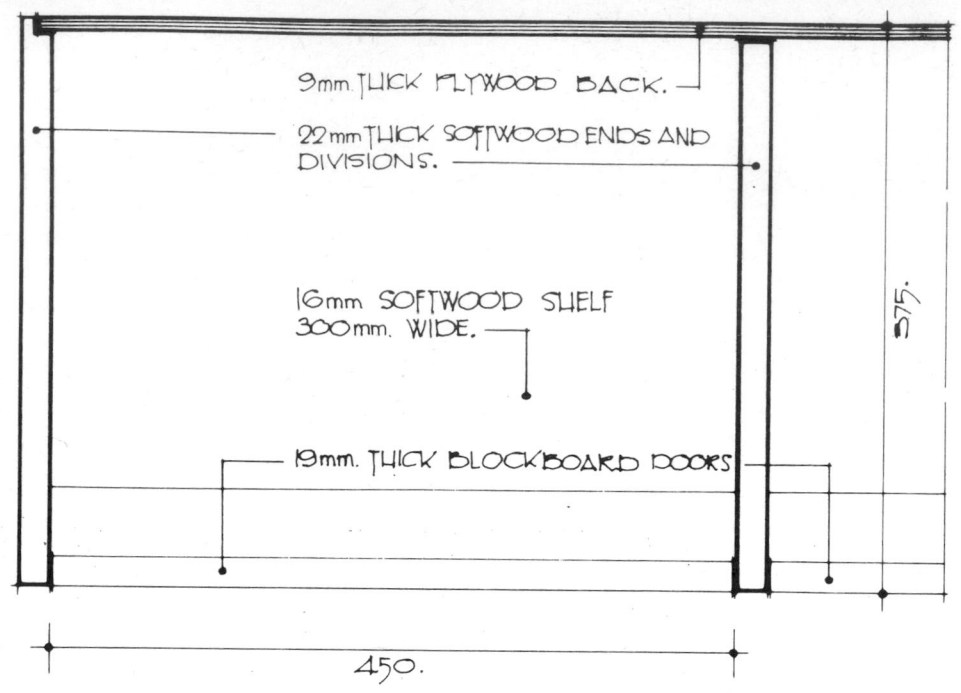

9mm THICK PLYWOOD BACK.

22mm THICK SOFTWOOD ENDS AND DIVISIONS.

16mm SOFTWOOD SHELF 300mm. WIDE.

19mm. THICK BLOCKBOARD DOORS

575.

450.

PLAN AT B. B.

22mm. THICK SOFTWOOD TOP.

22.

300.

19mm THICK BLOCKBOARD DOORS.

750.

16mm THICK SOFTWOOD SHELF.

22mm THICK SOFTWOOD DIVISIONS.

22 mm. THICK SOFTWOOD DIVISION.

22.

9mm PLYWOOD BACK.

22mm. THICK SOFTWOOD BOTTOM.

22.

100.

100×40mm THICK BEARERS.

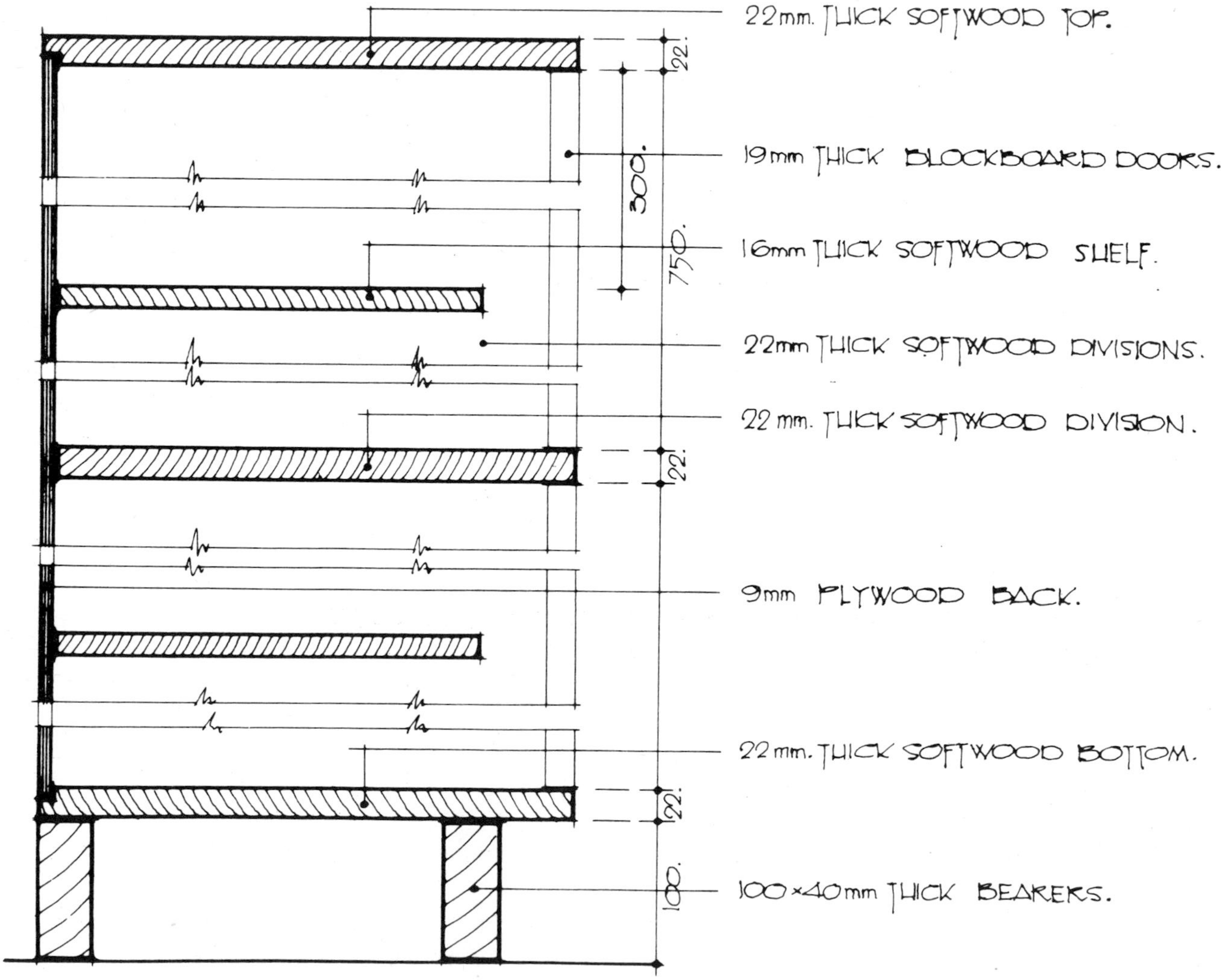

SECTION A. A.

SCALE. 1. : 5.

Plate 19
LOCKER UNIT

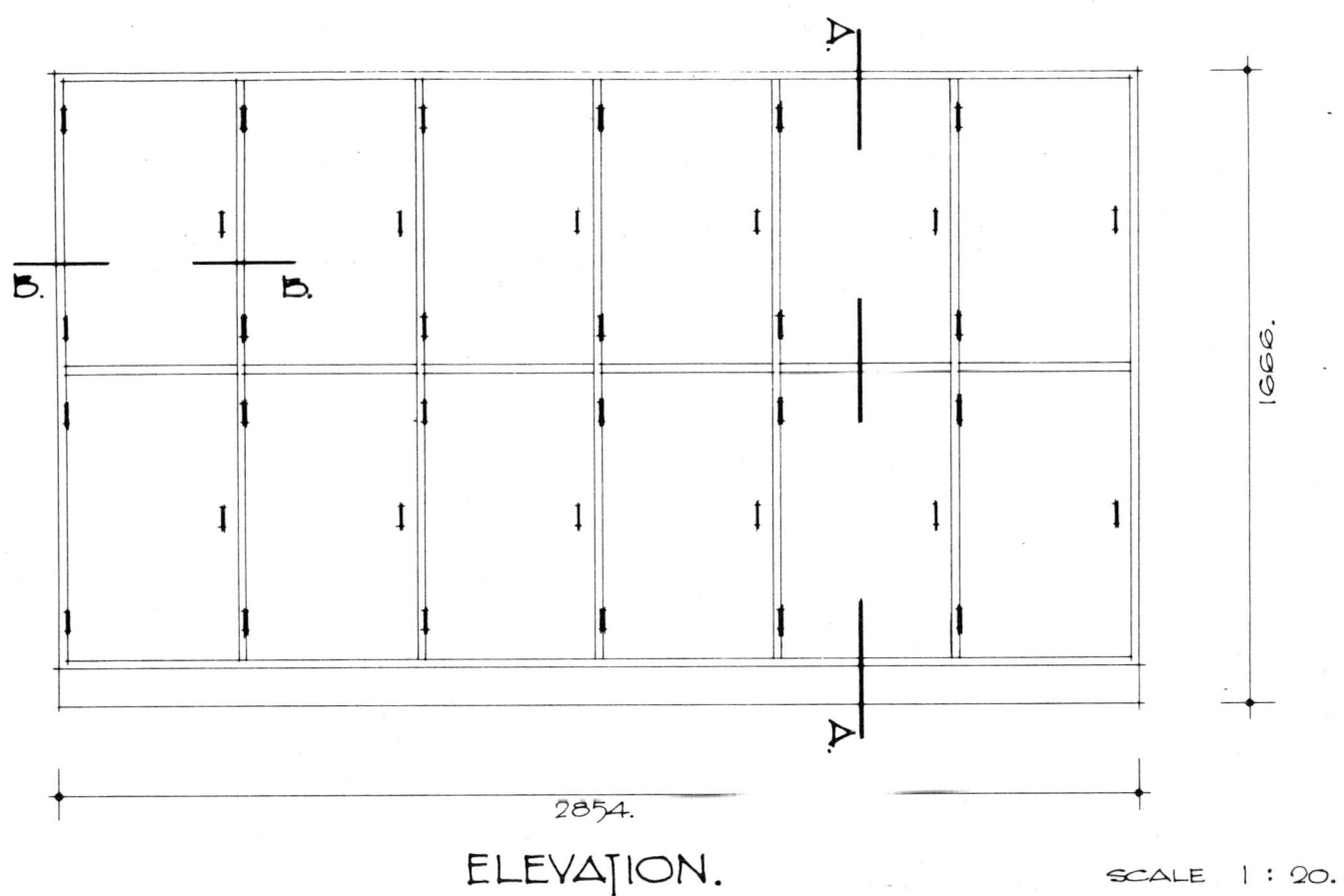

ELEVATION.

SCALE 1 : 20.

IRONMONGERY :-
EACH DOOR FITTED WITH PAIR 50mm. PRESSED STEEL BUTTS.
AND CUPBOARD LOCK. (P.W.A. REF. S.P. 12104).
K.P.S. 3.

Plate 20
UNFRAMED SECOND FIXINGS

S.M.M. RULES   SECTION A.
                 N.1.11.13.14.15.28.
                 V.1.2.3.4.

APPROACH:-   The S.M.M. Rules have been drafted on the
assumption that the majority of woodworking
is now a shop process using machinery rather
than a site craft process, hence the emphasis
on composite items and machine labours.
Unless otherwise stated or included as part
of a composite item, unframed second fixings
(site craft processes) are to be measured in
accordance with S.M.M. Rules N13 - N16.

PIPE CASING.

Ht	2.400
wktop	0.750
	1.650

S.M.M. N1.11.28.

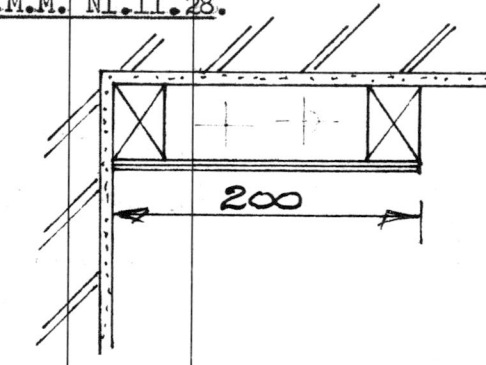

2/1.65   32 × 50 mm
Wrot sw batten
plugged to bkk.

1.65   3mm thk interior
qual birch faced
plywood pipe casg
n. e. 200 mm wide
scrd to sw.

&
K. P. S. ③ gen
wood surfs
n. e. 300 mm gth.
(new ark
1ntr

S.M.M. N15.
Kind of sheet   - plywood
Thickness   - 3mm
Joints   - not applicable
Fixing   - screws
Base structure - softwood
Classification - n. e. 200mm wide.

S.M.M. V1.2.3.4.

Isolated surface - abutting
walls plastered.

154

MENU BOARD.

S.M.M. N1.11.13.28.

	Batten 0.600
	2/0.050 0.100
2/1.00	0.500
2/0.50	

25x50 mm wrot sw batten plugged to bkk

S.M.M. N15.

	Batten 1.000
	2/0.050 0.100
	0.900
0.90	0.600
0.50	0.100
	0.500

25 mm thk "Tentest" insulation board to walls.

S.M.M. N13.

2/1.00
2/0.60

25 x 75 mm wrot sw. splayed architrave

S.M.M. V1.2.3.4.

	0.15
	0.75
	0.25
	0.25
	0.140

K.P.S. ③ wall mouldings n.e. 150 mm gth.   (new wk 1 wtr

S.M.M. V2.

Void exceeds 0.50m^2

1.00
0.60

Ddt
2cc emulsion plaster walls ab.

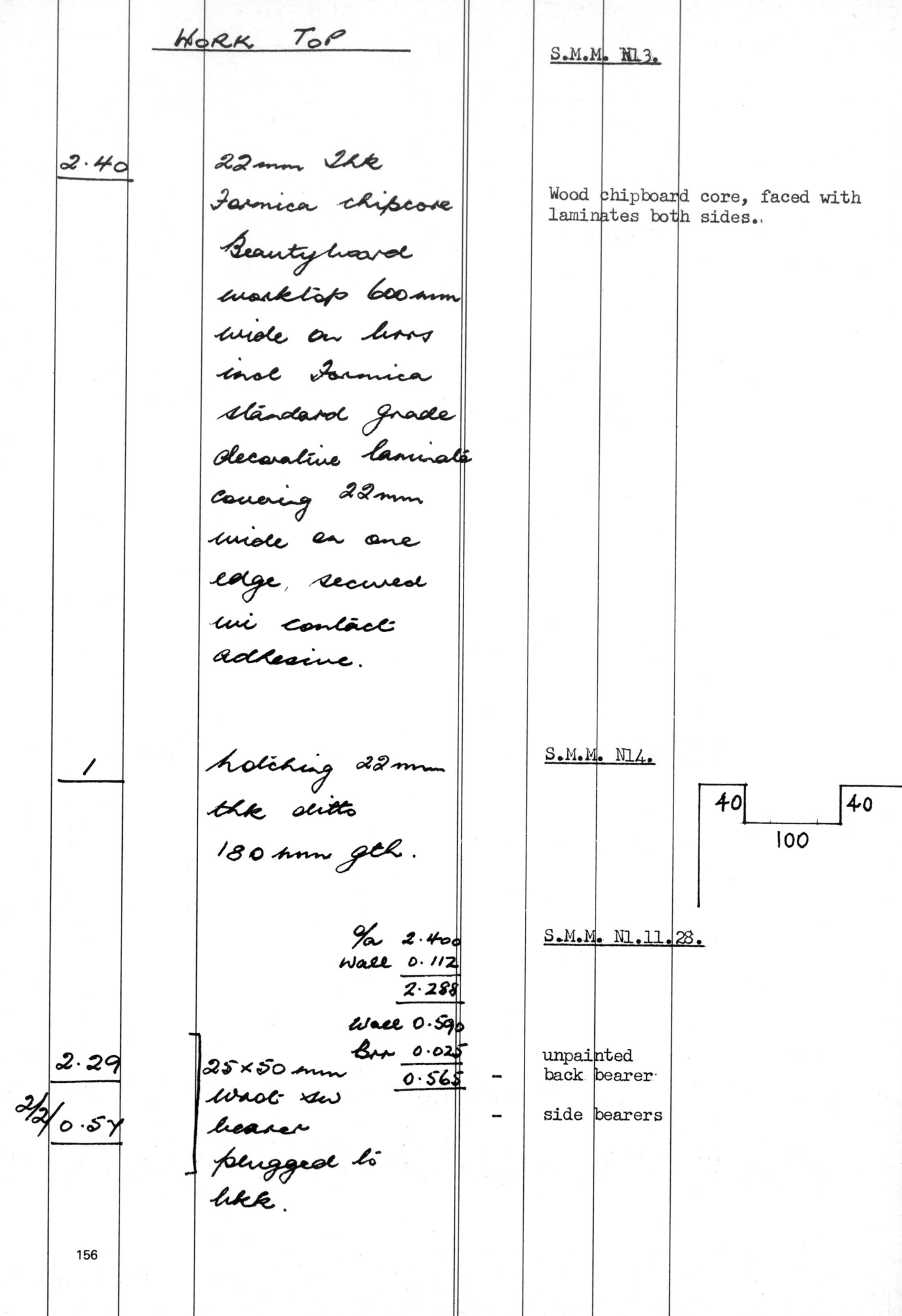

WORK TOP

S.M.M. N13.

2.40 | 22mm thk
Formica chipcore
Beautyboard
worktop 600mm
wide on brrs
incl Formica
standard grade
decorative laminate
covering 22mm
wide on one
edge, secured
wi contact
adhesive.

Wood chipboard core, faced with laminates both sides.

1 | notching 22mm
thk ditto
180mm gth.

S.M.M. N14.

```
40        40
    100
```

%a 2.400
Wall 0.112
      2.288

Wall 0.590
Brr 0.025
    0.565 —

S.M.M. N1.11.28.

unpainted
back bearer

2.29 | 25×50mm
woot sw
bearer

2/2/ 0.57 | plugged to
bkk.

— side bearers

SHELVING.

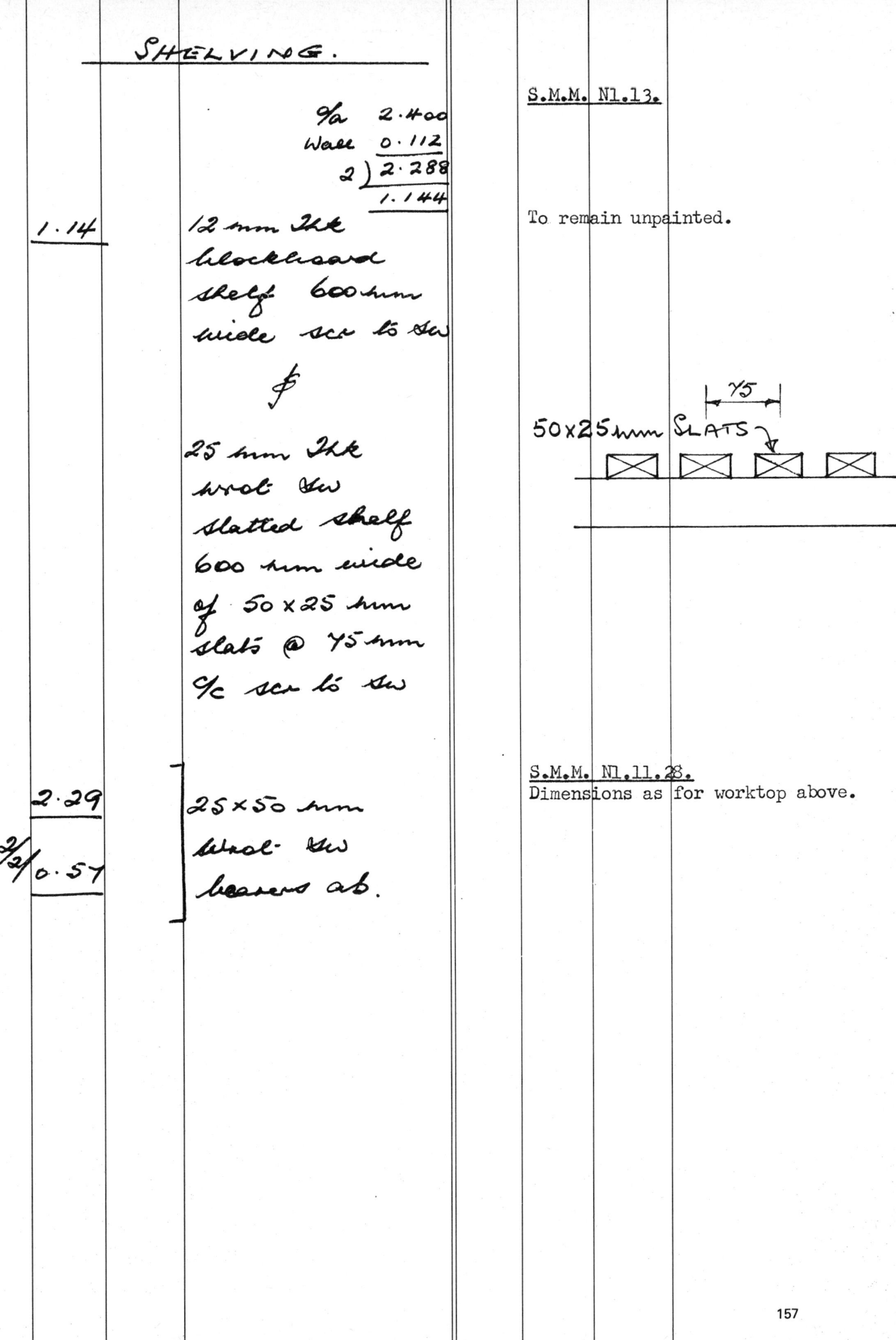

	%a	2.400
	Wall	0.112
	2)	2.288
		1.144

1.14   12 mm thk blockboard shelf 600 mm wide scr to sw

S.M.M. N1.13.

To remain unpainted.

&

25 mm thk wrot sw slatted shelf 600 mm wide of 50 x 25 mm slats @ 75 mm %c scr to sw

50x25mm SLATS

S.M.M. N1.11.28.
Dimensions as for worktop above.

2.29
²/₂/ 0.57   25 x 50 mm wrot sw bearers ab.

Plate 20
UNFRAMED SECOND FIXINGS

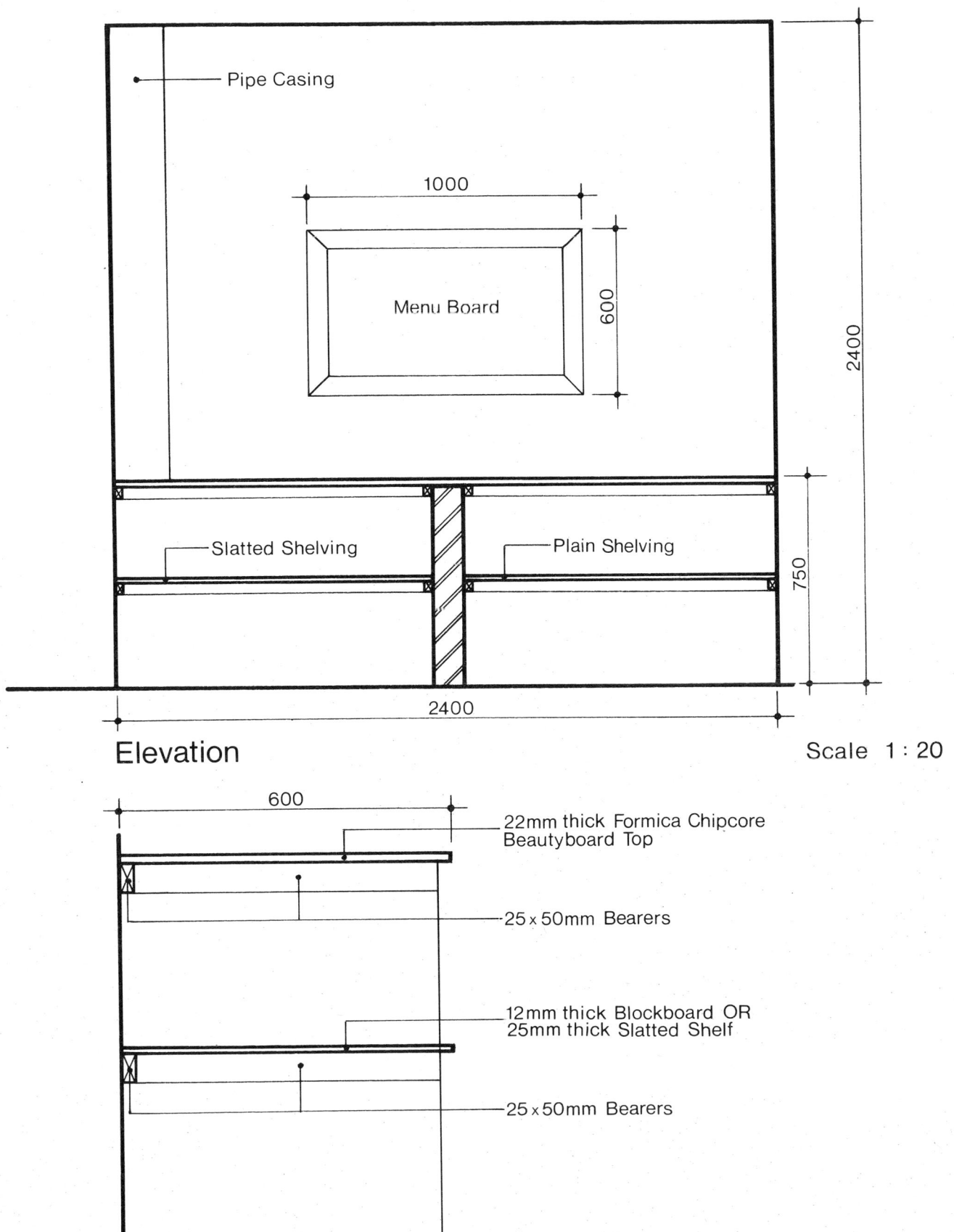

Pipe Casing

1000

Menu Board

600

2400

Slatted Shelving

Plain Shelving

750

2400

Elevation

Scale 1 : 20

600

22mm thick Formica Chipcore
Beautyboard Top

25 x 50mm Bearers

12mm thick Blockboard OR
25mm thick Slatted Shelf

25 x 50mm Bearers

Section

Scale 1 : 10

APPENDIX A

Abbreviations in common use

a.b.	as before		c.t.&b.	cut tooth and bond
a.d.	as described		chfd.	chamfered
addl.	additional		chy.	chimney
agg.	aggregate		clg.	ceiling
a.f.	after fixing		col.	column
ard.	around		cos.	course
art.	artificial		cpd.	cupboard
asb.	asbestos		conc.	concrete
asph.	asphalt		csk.	countersunk
avg.	average		ct.	cement
b & p.	bed & point		d/d	delivered
b.e.	both edges		Ddt.	deduct
b.f.	before fixing		d.h.	double hung
b.i.	build in		dia.	diameter
b.m.	birdsmouth		dist.	distemper
b.n.	bull nosed		D.P.C.	damp proof course
b.s.	both sides		D.P.M.	damp proof membrane
bal.	baluster		d.p.	distance piece
bast.	basement		dp.	deep
bdd.	bedded			
bdg.	boarding		E.M.L.	Expanded Metal Lathing
bk.	brick		E.O.	extra over
bkt.	bracket		ea.	each
bldg.	building		exc.	excavate
brd.	board		excn.	excavation
brrs.	bearers		extl.	external
bwk. or bkk.	brickwork		extg.	existing
b.o.e.	brick on edge			
B.S.	British Standard		F.A.I.	Fresh air inlet
B.S.C.	British Standard Channel		f.c.	fair cutting
B.S.E.A.	British Standard Equal Angle		f.f.	fair face
B.S.T.	British Standard Tee		f & b.	framed and braced
B.S.U.A.	British Standard		f.l.&b.	framed ledged and braced
	Unequal Angle		F.L.	floor level
B.S.U.B.	British Standard		fcgs.	facings
	Universal Beam		fdns.	foundations
B.M.A.	Bronze Metal Antique		fin.	finished
			fr.	frame
casmt.	casement		frd.	framed
c & f.	cut and fit		fwk.	formwork
c & p.	cut and pin		ftd.	fitted
c & s.	cups and screws			
c.b.	common bricks		G.F.	Ground floor
c.bwk.	common brickwork		G.I.	Galvanised Iron
cc.	centres		G.L.	Ground level
c.c.	curved cutting		g.m.	gunmetal
c.c.n.	close copper nailing		galv.	galvanised
C.E.	Cleaning Eye		grano.	granolithic
C.I.	Cast Iron		gth.	girth
clg.jst.	ceiling joist			
c.jtd.	close jointed		h.b.s.	herring bone strutting
C.P.	Chromium Plated		h.b.w.	half brick wall
c.o.e.	curved on elevation		hdb.	hardboard
c.o.p.	curved on plan		h.c.	hardcore
c.s.g.	clear sheet glass		hdg.jt.	heading joint

160

h.m.	hand made	r & s.	render and set
h.n.&w.	head nut and washer	r.f. & s.	render float and set
hoz.	horizontal	rad.	radius
H.P.	High Pressure	R.C.	Reinforced Concrete
h.r.	half round	r.c.	raking cutting
h.t.	hollow tile	rdd.	rounded
ht.	height	reinf.	reinforced or reinforcement
hwd.	hardwood	R.E.	rodding eye
h.w.	hollow wall	R.L.	reduced levels
inc.	including	r.l.jt.	red lead joint
intl.	internal	r.m.e.	returned mitred end
inv.	invert	r.o.j.	rake out joint
I.C.	Inspection Chamber	R.S.C.	Rolled steel channel
		R.S.J.	Rolled steel joist
Jap.	Japanned	R.W.H.	rainwater head
jst.	joist	R.W.P.	rainwater pipe
jt.	joint	reb.	rebated
jtd.	jointed	retd.	returned
		ro.	rough
K.P.S.	Knot, prime, stop		
		S.A.A.	Satin anodised aluminium
Lab.	labour	s.b.j.	soldered branch joint
l & b.	ledged & braced	s.d.	screw down
l.p.	large pipe	s.c.	stop cock
l & c.	level and compact	segtl.	segmental
		s.e.	stopped end
matl.	material	s.g.	salt glazed
m.g.	make good	s. jt.	soldered joint
M.H.	Manhole	s.l.	short length
M.S.	Mild steel	soff.	soffit
m.s.	measured separately	s.p.	small pipe
mis.	mitres	s.q.	small quantities
mo.	moulded	s.w.	stoneware
mort.	mortice	sk.	sunk
msd.	measured	sktg.	skirting
		sq.	square
n.e.	not exceeding	s & l.	spread and level
No. or Nr.	number	S & V.P.	soil and vent pipe
		stg.	starting
o/a	overall	Swd.	softwood
o.c.n.	open copper nailing		
o.s.	one side	T	tee
opg.	opening	T & G.	tongued and grooved
orgl.	original	t & r.	treads and risers
③	three oils	t.c.	terra cotta
		t.p.	turning piece
pbd.	plasterboard		
P.C.	Prime Cost	V.O.	Variation order
p & s.	plank and strut	V.P.	Vent pipe
plas.	plaster		
plasd.	plastered	wi.	with
p.m.	purpose made	w.g.	white glazed
p.o.	planted on	W.I.	Wrought Iron
pol.	polished	W.P.	waste pipe
pr.	pair	wdw.	window
Prov.	Provisional	wthd.	weathered
prep.	prepare		
pt.	point	X grain	cross grain
ptd.	pointed	X tgd.	cross tongued
ptg.	pointing		
ptn.	partition	mm	millimetre
P.V.A.	Polyvinyl Acetate	m	metre
P.V.C.	Polyvinyl Chloride	m²	square metre
pvg.	paving	m³	cubic metre
		kg	kilogramme

Joinery Labours

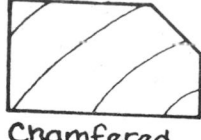

Chamfered

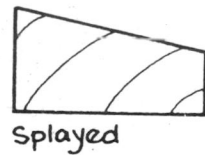

Splayed

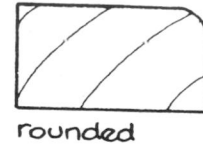

rounded

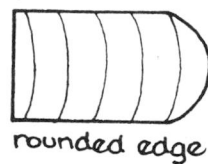

rounded edge

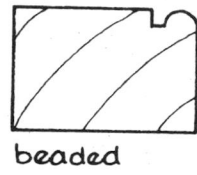

beaded

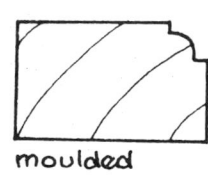

moulded

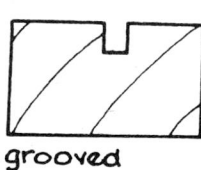

grooved

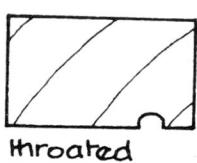

throated

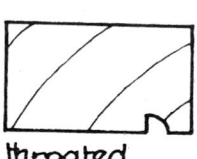

throated

N.B. A groove designed to prevent water from being drawn into the narrow space between two members is called a check-throat

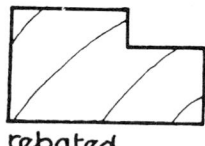

rebated

splay rebated

splay rebated

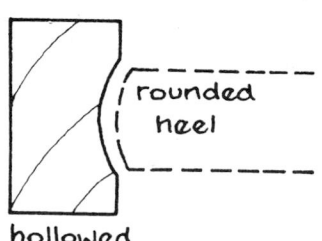

rounded heel

hollowed

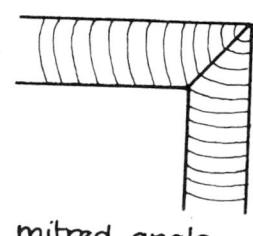

mitred angle

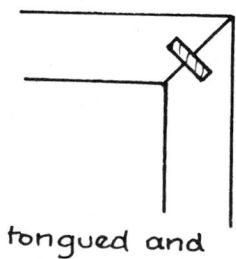

tongued and mitred angle

STOPPED LABOURS :–

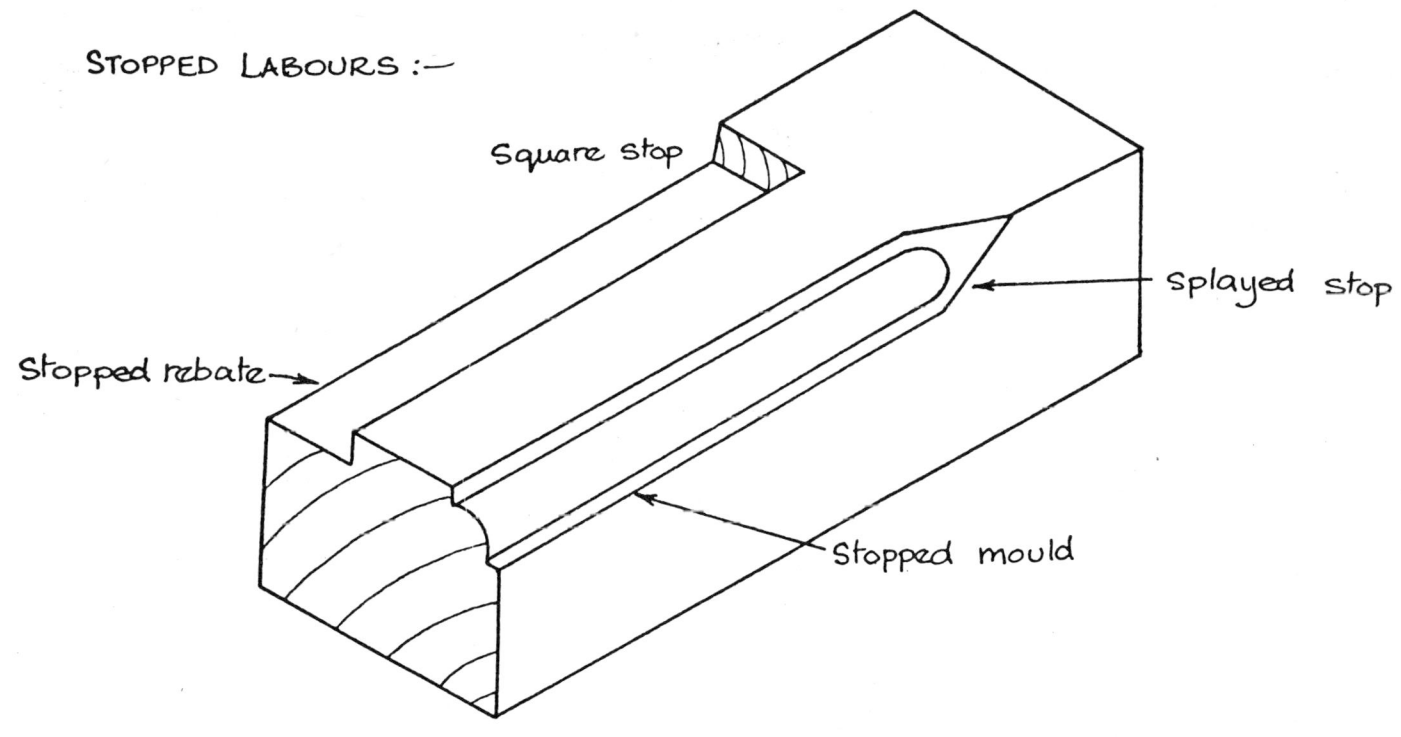

Square stop

splayed stop

Stopped rebate →

Stopped mould

WINDOW SILLS :–

Weathered

Sunk weathered

sunk weathered
and check throated

Sunk weathered
throated and
check throated

TABLE 11. EXAMPLE OF WINDOW SCHEDULE

TABLE 11. EXAMPLE OF WINDOW SCHEDULE (*contd.*)

DESCRIPTIONS	WINDOWS															
	1	2	3	4	5	6	7	8	9	10	11	12	13	14	15	16
SUB ASSEMBLY																
Type S1						+				+				+		
S2			+	+	+		+	+	+		+	+	+		+	+
S3		+														
S4	+															
WINDOWS																
Type W1						+				+				+		
W2			+	+	+		+	+	+		+	+	+		+	+
W3						+				+				+		
W4			+	+	+		+	+	+		+	+	+		+	+
W5						+				+				+		
W6			+	+	+		+	+	+		+	+	+		+	+
W7		+														
W8	+															
Hanging																
horiz slide			+	+	+		+	+	+		+	+	+		+	+
horiz pivot	+	+	+	+	+	+	+	+	+	+	+	+	+	+	+	+
fixed	+		+	+	+	+	+	+	+	+	+	+	+	+	+	+
Material																
anod al	+	+	+	+	+	+	+	+	+	+	+	+	+	+	+	+
SUB FRAMES																
Softwood																
890× 2690 mm						+				+				+		
1890× 2690 mm			+	+	+		+	+	+		+	+	+		+	+
890× 690 mm		+														
1890× 690 mm	+															
Transome w b p plywood																
100×50× 790 mm						+				+				+		
200×50× 790 mm						+				+				+		
100×50×1790 mm			+	+	+		+	+	+		+	+	+		+	+
200×50×1790 mm			+	+	+		+	+	+		+	+	+		+	+

sheet title WINDOW SCHEDULE	job no.	sheet no.
job title	CI/SfB (31) Xy	revision suffix A B
architect	page no. 2 of 3	date	drawn	checked

TABLE 11. EXAMPLE OF WINDOW SCHEDULE (*contd.*)

DESCRIPTIONS	WINDOWS																				
	1	2	3	4	5	6	7	8	9	10	11	12	13	14	15	16				
CLAZING																					
Thickness																					
3 mm (24oz)	+	+	1	1	1	1	1	1	1	1	1	1	1	1							
4 mm (32oz)			2	2	2	2	2	2	2	2	2	2									
5 mm (3/16 in)			3	3	3	3	3	3	3	3	3	3	3								
6 mm (1/4 in)														+	+	+					
Type																					
clear sheet			4	4	4	4	4	4	4	4	4	4	4								
obscured sheet	+	+																			
g w clear sheet			3	3	3	3	3	3	3	3	3	3	3								
g w pol plate			3	3	3									3	3	3					
polished plate														4	4	4					
Fixing																					
al beads 3 mm														+	+	+					
4 mm			3	3	3	3	3	3	3	3	3	3	3								
5 mm			2	2	2	2	2	2	2	2	2	2									
6 mm	+	+	1	1	1	1	1	1	1	1	1	1	1	1							

NOTES
1 indicates top light
2 " centre "
3 " bottom "
+ " all lights
4 " centre & top lights

sheet title WINDOW SCHEDULE	job no.	sheet no.
job title	CI/SfB (31) Ro	revision suffix A B
architect	page no. 3 of 3	date / drawn / checked

TABLE 12. EXAMPLE OF DOOR SCHEDULE

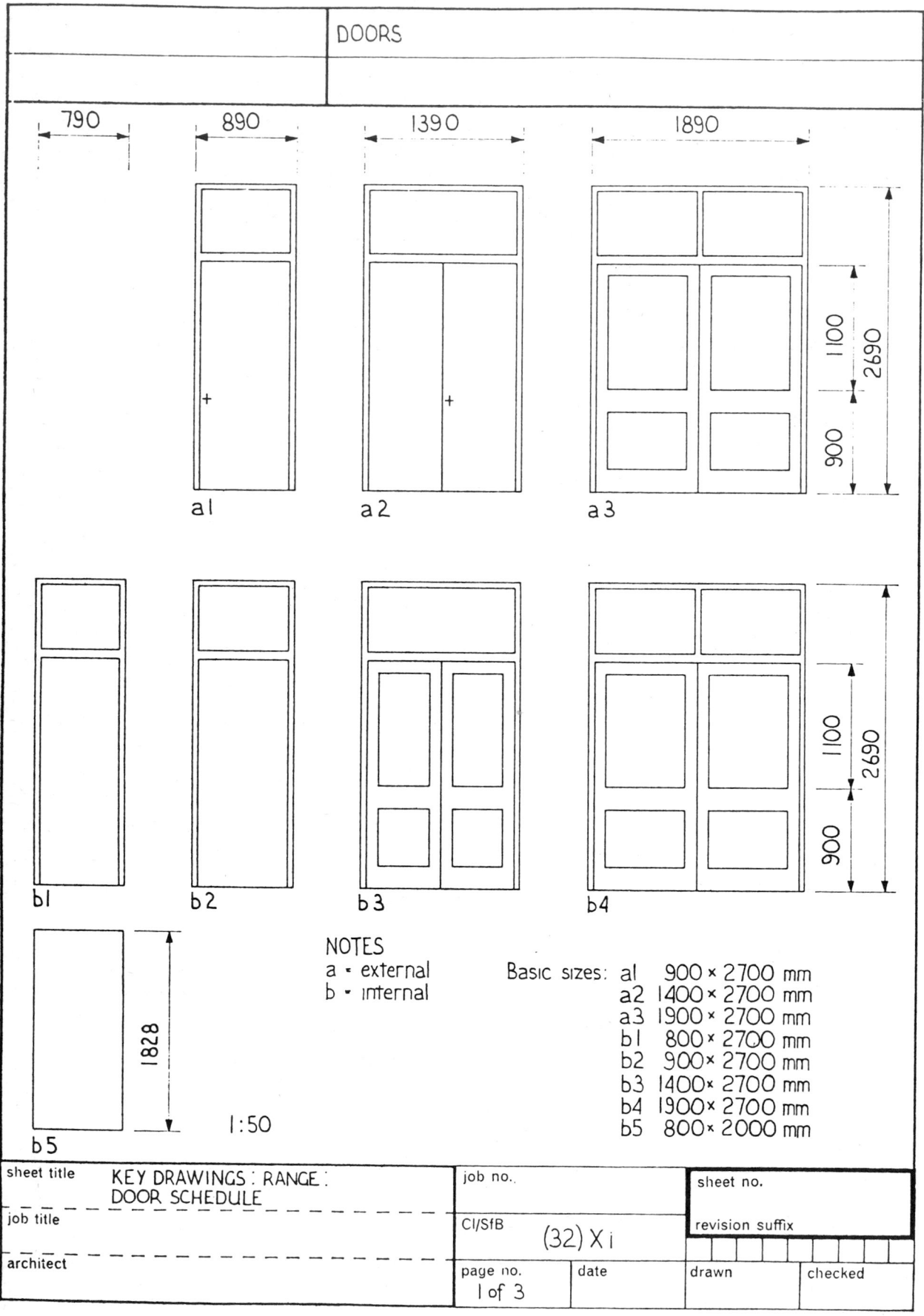

	DOORS

790   890   1390   1890

a1   a2   a3

b1   b2   b3   b4

b5   1:50

1828

NOTES
a = external
b = internal

Basic sizes:
a1 900 × 2700 mm
a2 1400 × 2700 mm
a3 1900 × 2700 mm
b1 800 × 2700 mm
b2 900 × 2700 mm
b3 1400 × 2700 mm
b4 1900 × 2700 mm
b5 800 × 2000 mm

sheet title	KEY DRAWINGS : RANGE : DOOR SCHEDULE	job no.		sheet no.	
job title		CI/SfB	(32) X i	revision suffix	
architect		page no. 1 of 3	date	drawn	checked

TABLE 12. EXAMPLE OF DOOR SCHEDULE (*contd.*)

DESCRIPTIONS	O/01	O/02	O/04	O/09	02/08	03/08	05/08	06/08	07/07a	08/09		
TYPE see key												
flush: solid core :ext q a1	+	+										
" " " :int q b1&2						+	+	+				
" semi-solid c: " " b5									+			
framed:glazed: a3				+								
" " b3					+							
" " b4										+		
" boarded a2			+									
LEAF SIZE 893×1990×57 mm				2						2		
643×1990×57 mm			2		2							
790×1990×44 mm	+	+					+	+				
690×1990×44 mm						+						
690×1828×44 mm									+			
STILE meeting												
square	+	+				+	+	+	+			
rounded			2	2	2					2		
MATERIALS												
hardwood			+	+	+					+		
plywood hwd face	+	+				+	+	+				
" paint q									+			
INFILLS												
623×1020×6 mm				2						2		
493×1020×6 mm					2							
623×535×6 mm				2						2		
493×535×6 mm					2							
Infill material												
clear plate glass					+							
gw plate glass				+						+		
Infill fixing												
hwd bead 15×20 mm				+	+					+		

NOTES
Door nos. digits refer to room nos. Prefix 'D/.....' in text.
O=outside. See clause 14·3
q = quality
gw = georgian wired

sheet title	DOOR SCHEDULE DOORS	job no.		sheet no.	
job title		CI/SfB (32) X i		revision suffix	
architect.		page no. 2 of 3	date	drawn	checked

TABLE 12. EXAMPLE OF DOOR SCHEDULE (*contd.*)

DESCRIPTIONS	O/01	O/02	O/04	O/09	O2/08	O3/08	O5/08	O6/08	O7/07a	O8/09		
FRAMES see key												
Type												
a1	+	+										
a2			+									
a3				+								
b1							+					
b2						+		+				
b3					+							
b4										+		
b5									+			
Shape												
rebated	+	+	+	+								
planted stop					+	+	+	+	+	+		
transome	+	+	+	+	+	+	+	+		+		
water bar	+	+	+	+								
fanlight mullion				+						+		
Materials												
hardwood			+							+		
softwood for paint	+	+	+		+	+	+	+	+			
FANLIGHTS												
Size												
1290 × 625 mm			+		+							
882 × 625 mm				2						2		
790 × 625 mm	+	+				+		+				
690 × 625 mm							+					
Materials												
9·5 mm (⅜ in) plasterboard							+					
5 mm (3/16 in) float glass		+			+	+				+		
5 mm (3/16 in) obsc. glass	+		+					+				
6 mm (¼ in) g.w. pol. plate				+								
Fixing												
hwd. beads 15 × 20 mm	+	+	+	+	+	· +	+	+	+	+		

sheet title	DOOR SCHEDULE FRAMES & FANLIGHTS	job no.		sheet no.	
job title		CI/SfB (32) X i		revision suffix	
architect		page no. 3 of 3	date	drawn	checked

TABLE 13. EXAMPLE OF IRONMONGERY SCHEDULE

DESCRIPTIONS	DOORS O/01	O/02	O/04	O/09	02/08	03/08	05/08	06/08	07/07a	08/09		
HANGING												
S A floor spring & pivot				2	2					2		
pair 100 mm (4in) brass butts	+	+	+									
" 100 mm (4in) steel "						+	+	+				
" " " " " falling butts									+			
FASTENING												
rim night latch	+S	+Z										
mortice latch & lock set						+Z	+S					
" dead lock			+	+	+					+		
master keyed			+	+	+	+Z	+S			+		
mortice latch	+S	+Z						+Z				
200mm recessed floor bolt			+	+	+					+		
" " " bolt to transome			+	+	+					+		
indicator bolt									+S			
panic bolt												
200mm S C kick plate	2	+		4	4					4		
100×200 mm S A A push plate				2	2					2		
100×200 mm S A A pull handle				2	2					2		
SAA lever furniture	+	+				+	+	+				
SAA pull bezel	+	+										
bell push				+								
S A A letter plate				+								
S A A coat hooks							+		+			
name plate 'BOYS CLOAKS'								+				
" " 'PRIVATE'							+					
overhead door closer	+S				+Z			+Z				

NOTES
Handing : s = clockwise
z = anti-clockwise
(subject to international agreement)

sheet title IRONMONGERY SCHEDULE	job no.	sheet no.
job title	CI/SfB (32) Xt 7	revision suffix
architect	page no. 1 of 1	date	drawn	checked